THE ART OF
ENAMELING

THE ART OF
ENAMELING

Techniques

Projects

Inspiration

LINDA DARTY

Asheville

editors: **Marthe Le Van, Deborah Morgenthal**
art director: **Kathleen Holmes**
cover design: **Barbara Zaretsky**
assistant editors: **Rebecca Guthrie, Nathalie Mornu**
assistant art director: **Shannon Yokeley**
editorial assistance: **Delores Gosnell**
art interns: **Melanie Cooper, Jason Thompson**
photographers: **Evan Bracken**, process shots;
keithwright.com, projects and samples
illustrators: **Orrin Lundgren, Sharon Massey**
proofreader: **Sherry Hames**

LARK CRAFTS

An Imprint of Sterling Publishing
387 Park Avenue South
New York, NY 10016

First Paperback Edition 2006
Text © 2004, Linda Darty
Photography © 2004, Lark Crafts, an Imprint of Sterling Publishing Co.,
Inc.; unless otherwise specified
Illustrations © 2004, Lark Crafts, an Imprint of Sterling Publishing Co.,
Inc.; unless otherwise specified

ISBN 978-1-57990-507-1 (hardcover) 978-1-57990-954-3 (paperback)

The Library of Congress has cataloged the hardcover edition as follows:

Darty, Linda.
    The art of enameling : techniques, projects, inspiration /
Linda Darty.
      p. cm.
    Includes bibliographical references and index.
    ISBN 1-57990-507-2
    1. Enamel and enameling—Technique.   I. Title
NK5000.D27 2004
738.4—dc22
                                                    2004005540

Distributed in Canada by Sterling Publishing
c/o Canadian Manda Group, 165 Dufferin Street
Toronto, Ontario, Canada M6K 3H6
Distributed in the United Kingdom by GMC Distribution Services
Castle Place, 166 High Street, Lewes, East Sussex, England BN7 1XU
Distributed in Australia by Capricorn Link (Australia) Pty. Ltd.
P.O. Box 704, Windsor, NSW 2756, Australia

For information about custom editions, special sales, and premium
and corporate purchases, please contact Sterling Special Sales at
800-805-5489 or specialsales@sterlingpublishing.com.

Email academic@larkbooks.com for information about desk and exami-
nation copies. The complete policy can be found at larkcrafts.com.

Every effort has been made to ensure that all the information
in this book is accurate. However, due to differing conditions, tools,
and individual skills, the publisher cannot be responsible for any
injuries, losses, and other damages that may result from the use
of the information in this book.

Manufactured in China

20   19   18   17   16   15   14   13   12

larkcrafts.com

Front cover: Kimberly Keyworth *Flower Vase*, 2003.
2½ x 1½ in. (6.4 x 3.2 cm). Sterling silver, 22-karat
gold, enamel; fabricated, torch fired. Photo by
George Post.

Back cover, top: Linda Darty *Outside In: Cup*, 2002;
bottom left: Tim Lazure *Fold-formed Bowl*, 2003; bottom right: Adrienne Grafton *Cloisonné Brooch*, 2003.

Title page: Linda Darty *Garden Brooches*, 2003.
4 x 2½ x ½ in. (10.2 x 6.4 x 1.3 cm). Fine silver,
enamel, gemstones, 14-karat gold. Photo by artist.

Top left and spine: Rebekah Laskin *Brooch*, 2000.

Top right: Jacqueline Ryan *Pendant*, 1996. 2¼ in.
(5.7 cm). 18-karat gold, enamel. Photo by Giovanni
Corvaja.

Center: Marjorie Simon *Red Fluffy Necklace*, 2002.

Bottom: Kathleen Browne *Masked*, 2002.

Contents

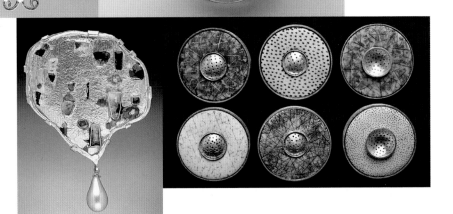

Top left: Jamie Bennett *Chadour Brooch #6,* 2000. 2½ x 1 x ¾ in. (6.4 x 2.5 x 1.9 cm). 18-karat gold, copper, enamel. Photo by Dean Powell.

Top right: Harlan Butt *Horizons: Big Cypress #1,* 2004.

Bottom left: Yoshiko Yamamoto *Brooch: After Gustav Klimt I,* 2002.

Bottom right: Jessica Turrell *Sift Series Brooches,* 2002.

Introduction

I AM ESPECIALLY PROUD to be writing this book about contemporary enameling techniques, not only because enameling is a passionate interest of mine but also because I believe it's a material that many have yet to discover.

I first learned about enameling through my interest in ceramics, discovering that the two shared many of the same materials. One evening, nearly 30 years ago, while working at Penland School of Crafts, I walked from the ceramics studio through the enameling area, and saw Barbara Mail, the instructor, using underglaze pencils. I was familiar with these pencils because I used them in my ceramics work to draw on bisque-fired porcelain. When I questioned Barbara, she informed me that it was all the same—glass on clay or glass on metal. Her remark hit me like a lightening bolt! I marveled that the enamel was detailed and glassy after only two or three minutes in the kiln, and that the layers of color could be applied and fired many times with immediate results.

Linda Darty *Garden Brooches: Spring*, 2003. 4 x 2½ x ½ in. (10.2 x 6.4 x 1.3 cm) each. Fine silver, enamel, gemstones, 14-karat gold. Photo by artist.

During the seven years I worked at Penland, I studied with dozens of metal and enameling teachers. The more I learned, the more I found I didn't know. Each session brought new students, and after my work day, they graciously caught me up on whatever they had been learning. When Bill Helwig, my first enameling teacher, showed me how to paint with enamel, the powdered glass felt familiar and the activity was reminiscent of my childhood experiences with paints and paper. I was at ease with enamel, and I loved the transparency and depth in the glass. I studied often with Bill, Jamie Bennett, Martha Banyas, Mel Someroski and many other excellent artists. I sifted and painted, made wall panels, objects, and jewelry. I worked with whatever technique the teacher was teaching, and moved quickly and easily from one to the next. I learned to fire in a kiln and with a torch, and to apply enamel in numerous ways. I knew nothing about metalworking until eventually my friends showed me that it wasn't as hard as it looked.

I kept detailed notes from each teacher, and asked hundreds of questions. Truly it is those notes and those questions that are the foundation for this book, and I'm proud and pleased that what so many gifted teachers shared with me, I can now pass on to you.

I'm fortunate now to enjoy a career not only as a maker but also as an educator, a profession that allows me to share my passion for enameling. It's a great pleasure to see students do such wonderful things with glass on metal, using it not only with great respect for its history and tradition, but also in innovative ways that express their unique vision as artists. I have learned that my role as their teacher is to demystify enameling; to teach it as merely another tool for expressing ideas in metal, for adding rich color, texture, or imagery to their work. I teach my students the techniques in the order they're presented in this book, and once they master them, I encourage them to develop experimental approaches. By the second semester I find that students may ask me questions about how they should join or make something when planning an enamel project, but rarely do they question whether it's possible. To enjoy this freedom, you must learn a few basic techniques and become familiar with the many diverse effects that can be achieved with glass on metal.

If you're new to enameling you should begin by reading the Enameling Fundamentals section of this book, which presents basic information that relates to all enameling techniques. The next section, Enameling Techniques, explains different ways to enamel; I have intentionally kept the procedure descriptions brief so that there is more room for pictures that demonstrate the processes. Historical notes are included, simply because I find this history so fascinating, and also because it's inspiring to see and learn from these works that were created long ago. Each chapter includes photographs of contemporary enamel work. I've also included the work of students, many of whom have studied enameling only a short time, emphatically demonstrating that most techniques are not difficult. The final section offers projects with detailed instructions, explaining procedures so that you can develop your own personal work with an understanding of the steps involved.

I believe that the more technological our world becomes, the more important it is that we stay connected to creative activities such as enameling. Making art involves our hearts and our hands, an important endeavor that modern culture seems to value less and less. It is my hope that you'll become seduced and enthralled by the stunning colors and the infinite possibilities enameling on metal offers, and that this book will help you find great satisfaction, not only in what you make, but in the very act of making.

Linda Darty *Garden Candlestick,* 1999. 4 x 3 x 3 in. (10.2 x 7.6 x 7.6 cm). Sterling silver, enamel. Photo by Henry Stindt.

Linda Darty *Floating Heart Nymphoides Peltara Candlesticks,* 2003. 8 x 6 x 6½ in. (20.3 x 15.2 x 16.5 cm). Enamel, copper, silver. Photo by Robert Diamante.

Enameling Fundamentals

Sarah Perkins *Cactus Container,* 2004. 6 x 4 in. (15.2 x 10.2 cm). Fine silver, enamel. Photo by artist.

IN THIS CHAPTER, you'll learn basic enameling skills including the different types of enamels and their characteristics, what metals you can use for enameling, how to set up an enameling studio, and what tools and equipment you'll need. You'll understand how to clean the metal, prepare and apply the enamel, fire a piece, and clean between firings. You'll find out how to remove enamel if you make a mistake, don't like a color, or discover unwanted pits or impurities in the glass. You'll learn to finish the glass, an important part of the enameling process that allows you to control surface textures. You'll be able to create enamels that are highly glossy, silky satin, frosty matte, and even sugar- and pebble-textured. I've included a section on ways to set a finished enamel in metal, so you can consider your options before you design a piece. There is also a section that gives you a simple method for testing enamel colors, so you'll be well-prepared to work with shading and layering colors as required in the second section of this book, Enameling Techniques. If you're new to enameling, it's important to read and understand this section completely. If you already feel comfortable with the basics, you may want to glance through this information to refresh your memory, and then move on to the second section and begin exploring various techniques.

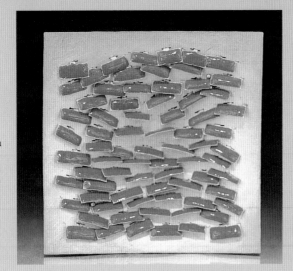
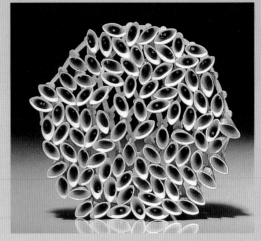

Enamel & Enameling

Vitreous enamels were useful to ancient craftspeople because they could serve as substitutes for precious gemstones in jewelry and ceremonial objects. The earliest colors were meant to replicate specific stones: cobalt glass replaced rare lapis lazuli, opaque blue-green represented turquoise, and a reddish brown color could substitute for garnet or carnelian. Even before the earliest enamels were created, ancient artists had a fascination with glass; the Mycenaean Greeks used to string pieces of glass pierced with holes to create necklaces and other jewelry objects, and stone molds to cast molten glass have been found in Greece. By 2,000 BC, craftspeople had become adept at working with bronze and were able to solder gold and silver, therefore setting the stage for the first developments in enameling technique.

THE PROPERTIES OF ENAMEL

Manufacturers of enamel today carefully test the properties of the glass they make. They control the composition so that it bonds to different metals and has different physical properties related to its *softening point*, (which indicates the temperature when the enamel will begin to move or flow) *fusion flow* (which indicates the rate or speed that it will flow at a particular temperature) and *coefficient of expansion* (the increase in the length, or volume of the enamel during heating, the "fit" or compatibility with the metal).

Though as artists we usually purchase the enamel without regard to this information, it's helpful to understand these physical properties so that as you become familiar with the colors you use, you'll know why they react to layering and firing in certain ways. Sometimes for example, a color may be listed from the supplier as Hard, Medium, or Soft. This relates to the softening point of the glass, and how hot it should be fired. (Enamel does not melt; it softens and flows.) Not the same as the enamel's softening point, the fusion flow is more descriptive of how quickly the enamel spreads and smooths out. Two different enamels might soften at the same temperature, but one could flow and smooth out more easily than the other one, a fact you might be concerned with if you work on vertical surfaces. This helps us understand why some enamels break through the colors that are fired on top of them. An enamel with a high fusion flow (which flows out and smooths easily) and a lower softening point (which fires at a lower temperature) would work best if used on top of an enamel which fires at a higher temperature (that has a higher softening point and flows less quickly.) If the enamel that fires at the higher temperature is used on top, the softer color will flow and break through the harder color. Not all enamels are marked or labeled to indicate their softening points, fusions flows, and coefficients of expansion, but as you use colors you'll soon learn about these properties, even if your understanding of how they fire is more intuitive than analytical.

HOW ENAMEL IS MADE

To make enamel for metal a combination of raw minerals (such as silica, soda ash, and potassium nitrate) are mixed according to each enamel formula and put into a preheated crucible, similar to the way glass blowers make raw

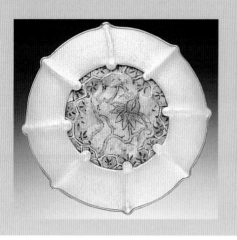

Left: Sydney Scherr *Floating*, 2001. 3¾ x 3½ x ½ in. (9.5 x 8.9 x 1.3 cm). 24-karat gold, 22-karat gold, 18-karat gold, fine silver, opalescent enamel, blue pearl, opal beads; fabricated, hand engraved, cloisonné. Photo by Seth Tice-Lewis.

Far left: Jamie Bennett *Florilegia Brooch*, 2003. 2¾ x ½ in. (6.9 x 1.3 cm). 20-karat gold, enamel. Photo by Dean Powell.

"batch" for glass blowing. Changing the proportions of the minerals in the formula will change the properties of the enamel (its softening point, fusion flow, and coefficient of expansion). Ceramic pigments can be added to the glass to control color; for example cobalt makes blue, chrome green, cadmium is in yellows. (Though many artists often say it is "oxides" that combine with glass to create color, some of the organic pigments contain oxygen, but others do not. The oxygen in cobalt oxide, for instance, has nothing to do with producing the color.)

Insoluble crystals can be added to decrease the transparency of the color and make it more opaque. Once the glass is heated to the

A

correct temperature for the correct amount of time it takes to melt the minerals, a viscous liquid is created. This glass is removed from the crucible, air cooled on an iron plate or in water, then shattered into lump or flake enamel (see photo A). The enamel is then ground into granules that are measured by passing the powdered glass through a screen with a standardized number of openings in 1 linear inch (2.5 linear cm). The more openings in the screen per inch or centimeter, the smaller the granules that pass through it. This screening determines the enamel mesh size and most commercial enamels are sold as 80 mesh.

TYPES OF ENAMEL

Depending on the composition of the glass, enamel can be clear, transparent, opaque, or opalescent. Opaque colors completely block what's beneath them from view. This applies to both the surface of the metal and other enamel layers. Opalescent enamel is a translucent glass that's not completely clear. Its milky appearance is similar to that of an opal gemstone. Transparent enamel can be either completely clear or a color.

LEAD-BEARING & LEAD-FREE ENAMELS

Lead-bearing and lead-free colors are available to enamel artists. I use them both, primarily basing my choice on the color or technique I want to use. Here are some guidelines to help you make an informed decision.

• Lead-free colors are much safer to use than lead-bearing colors.

• Lead-bearing enamels are available in a much greater range of colors. This is because more companies manufacture lead-bearing products.

• Lead-bearing enamels are safer to use when you apply them wet as shown above. By mixing wet enamels in a watercolor tray, you're less likely to breathe sifted dust containing lead. Because there are more colors, and

Right: Marilyn Druin
Golden Sampler, 2001. 2⅓ x 2 ⅓ in. (5.9 x 5.9 cm). 24-karat yellow gold, 18-karat yellow gold, fine silver, enamel; cloisonné, guilloché, basse taille. Photo by Bob Barrett.

Far right: Felicia Szorad
Canteen Bag, 2001. 10 x 7 x 5 in. (25.4 x 17.8 x 12.7 cm). Sterling silver, bronze, copper, enamel. Photo by Taylor Dabney.

because they are safer when used wet, I often choose lead bearing colors for cloisonné or champlevé enameling techniques. You could, however, use lead-free colors just as easily.

• Using lead-free colors when sifting dry enamels reduces your exposure to toxic lead. However, both lead-free and lead-bearing glasses contain other metal oxides and pigments that are toxic to inhale. Always wear a safety mask when sifting dry colors.

• Lead-free enamel colors are more acid-resistant. You should use them when you intend to clean firescale off an uncoated copper surface with an acid solution. (For more information on this technique, see page 25.)

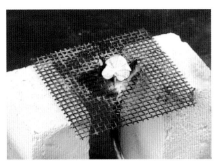

• Always choose lead-free colors when you plan to torch-fire enamel. This decreases your risk of breathing lead fumes emitted during torch-firing, and when affected by the carbon in the torch flame, lead-free colors don't discolor as easily as lead-bearing colors.

B

As explained earlier, when enamel is heated it expands and the measurement of this change is called its *coefficient of expansion*. Due to the different coefficients of expansion of lead-bearing and lead-free enamels, they seem most stable in combination when the lead-bearing color is used on top of the lead-free color. For example, a small amount of a beautiful red lead-bearing

color might serve well as a final inlayed area in an overall lead-free enameled piece. The opposite combination is also possible, though sometimes when the lead-free color is applied on top, the lead-bearing bottom color breaks through and creates a texture (see photo B). It is interesing to know that both lead-bearing and lead-free enamels were used together in Limoge, France, in the 16th century. These two types of enamel still work together, even if "sandwiched" in layers. A single grain of lead enamel on a lead-free surface, however, can cause a pit to occur.

My suggestion to any beginning enamelist is to start with a limited number of transparent and opaque colors, depending on your preference. You can use a lead-free palette, a lead-bearing palette, or, perhaps even a combination of the two. Experiment with layering just a few colors in many different ways. Learn what combinations work best for you and create effects you enjoy. Although over the years I've collected shelves of enamel colors in my studio, I find that I usually work with the same colors over and over again, adding new ones only occasionally.

Enameling Materials

One of the great advantages contemporary enamelists enjoy is the opportunity to purchase enamel in many forms. In comparison to the medieval artists who made the glass for each enamel in hand-built wood-heated furnaces, our options are incredibly convenient, offering many more diverse and straightforward ways to work with glass on metal. As you explore the Enameling Techniques section of this book, you'll learn how to use each of these materials.

POWDERED ENAMEL

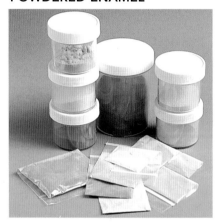

Most commercial powdered enamels are sold as 80-mesh. You can make or purchase your own 60-mesh, 80-mesh, 100-mesh, 200-mesh and 325-mesh screens and sift enamel powder into smaller particle sizes, separating it for different applications.

PAINTING ENAMELS

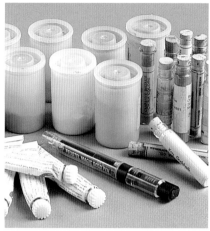

With particles typically ground smaller than 325-mesh, painting enamels are even finer than liquid enamels. You can purchase painting enamels premixed with oil in tubes, or you can buy them dry and mix them with oil or water as needed. When mixed, painting enamels are very similar to artist's oil paints, and you can apply them in layers over fired enameled surfaces.

WATERCOLOR AND ACRYLIC ENAMELS

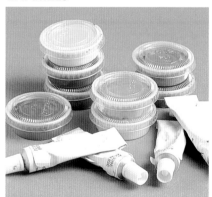

These enamels are easy to work with for those who like to paint and draw. Water-based enamel paints are sold in tubes like acrylics and in pans like watercolors.

ADDITIONAL MATERIALS

More interesting materials for enameling include: underglaze pencils, chalks, ceramic pigments, and crayons (photo A); lumps, chips, and threads (photo B); and decal papers (photo C).

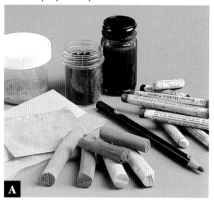

LIQUID FORM WATER-BASED ENAMEL

You can purchase lead-free liquid enamel, premixed with water or in dry form. Liquid enamel consists of glass and additives to control the drain and set time. These components are ground together until all but a small percentage passes through a 200-mesh screen.

—— Vitreous enamels are composed of glass and inorganic pigments, such as cobalt, used to produce color. The mineral composition of the glass gives it different properties related to its softening point, flow rate, and fit to the metal.

—— Powdered enamel is ground to varying degrees and designated by mesh size. For many sifting applications, 80-mesh and 150-mesh powders are practical sizes. These particles have a similar consistency to sugar or table salt. If needed, you can sift finer enamel by using different size mesh screens. For painting, 200-mesh and 325-mesh powders are better choices.

—— You can purchase transparent, opaque, or opalescent enamel colors as well as clear enamel, which is commonly called *flux*.

—— Lead-free enamels are safer to use and more acid-resistant. Lead-bearing enamels are available in a wider variety of colors. All enamels, both lead-free and lead-bearing, contain ingredients that can be hazardous when inhaled.

—— A lead-bearing color fired on top of a lead-free color is compatible more often than a lead-free color fired on top of a lead-bearing color. You should try the reverse application, as it sometimes yields interesting textural effects.

HISTORICAL HIGHLIGHT

Theophilus Presbyter, a 12th-century German monk, described preparing enamel and testing its properties in his medieval treatise, *On Divers Arts:*

> ...take all the kinds of glass you have prepared for this work and breaking off a little from each piece put all fragments at the same time on a single sheet of copper, each fragment, however, by itself. Then put it into the fire and build up coals around and above it. Then while you are blowing observe carefully whether the fragments melt evenly; if so, use them all; but if any fragment is more resistant, lay aside by itself the stock that it represents. Now take all the pieces of tested glass and put them one at a time in the fire and when each one becomes red-hot throw it into a copper pot containing water, and it will immediately burst into tiny fragments. Quickly crush these fragments with a pestle until they are fine. Wash them and put them in a clean shell and cover with a linen cloth. Prepare each color separately in this way.

Etienne Delaune *Goldsmith's Workshop*, 1576. $3\frac{1}{10}$ x $4\frac{7}{10}$ in. (8 x 12 cm). Engraving. © Copyright The British Museum.

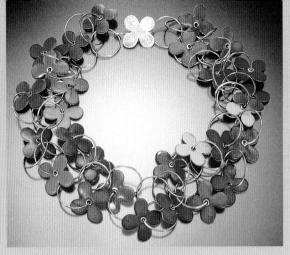

Left: **Marjorie Simon** *Red Fluffy Necklace,* 2002. 17 x 1 in. (43.2 x 2.5 cm). Copper, sterling silver, enamel. Photo by Ralph Gabriner.

Far left: Jessica Calderwood *Maintaining Lift,* 2000. 18 x 18 in. (45.7 x 45.7 cm). Enamel, steel. Photo by Michael Cirelli.

Setting Up an Enameling Studio

You can enamel in a modest space with a minimum amount of equipment. If you have a larger studio, you can create two distinct areas for metalworking and enameling. This separation makes it easier to keep the enameling area clean and free of metal dust, but it's a luxury, not a necessity. Design an area that works well for you, bearing in mind that it must be away from living, dining, and food preparation spaces.

STUDIO SPACE

• A clean work area

Good housekeeping is essential. If you work in metal, you'll want a separate clean table for enameling or at least the discipline to clean up your metalworking bench with a damp cloth on the days you plan to enamel. (Intentionally sprinkling metal filings or firescale into enamel to create special effects is different from finding a deep unexpected pit in the fired glass due to stray filings, steel wool, or dirt.)

• A sink with at least cold running water

• Good lighting

You can easily adjust the position of flexible clip-on lights over your work.

TOOLS & SUPPLIES

The basic tools and materials required for all types of enameling are listed here. You'll learn about more specialized equipment later in the book, as it becomes applicable to a particular technique.

• A heat source

To fuse enamel, you'll need access to a kiln that reaches at least 1500° F (815° C) or a propane or an acetylene torch with medium to large tips. (Refer to pages 17–19 for more information on kilns and torches.)

• A heat-proof surface, such as firebrick

You'll need somewhere to place hot items after firing.

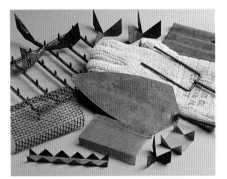

• Trivets and screens

Whether using a torch or kiln, you'll need well-constructed firing supports on which to place your work.

• A sheet of mica

Sheet mica makes a good support for pieces that are otherwise difficult to fire on a trivet, or for plique-á-jour enamels.

• A firing fork or trowel

You'll use a fork or trowel to place an enamel into a hot kiln.

• Fireproof gloves

Wearing fireproof gloves when you kiln-fire enamel is a vital safety precaution, especially if a hot piece falls and needs to be retrieved!

• A good dust mask

Make sure this safety device is approved for blocking fine airborne particles.

Right: Felicia Szorad *Parts for Eventual Jewelry Pieces*, 2001. 1 x 2 x ¼ in. (2.5 x 5 x .6 cm). Copper, enamel; torch fired. Photo by Taylor Dabney.

Far right: Sandra Zilker *Points/Petals Pin*, 2003. 4¼ x 4¼ x ½ in. (10.8 x 10.8 x 1.3 cm). Sterling, copper, enamel; torch fired. Photo by Jack B. Zilker

• A table shear, hand shear, or jeweler's saw
Unless you only want to enamel on pre-cut metal shapes, you'll need some sort of cutting tool.

• A flexible-shaft machine, drill press, or hand drill
Use these machines to make function holes in metal, such as for jewelry findings, or to create decorative elements, such as pierced metalwork.

• A fiberglass brush
You'll use a fiberglass brush to clean the enamel surface between firings.

• A 6-inch (15.2 cm) half-round medium-cut or fine-cut file, or an alundum stone
You can use any of these tools to clean the edges of copper between firings.

• Wet/dry carborundum paper
Keep a good stock of this finishing paper on hand in 320-, 400-, and 600-grits.

• Alundum stone or diamond sanding sticks
These tools abrade enameled surfaces. You can purchase them from enamel supply companies.

• Small glass or plastic containers with tight-fitting lids
You'll need a variety of containers for washing and storing enamels.

• A bucket, container, or coffee filter and jar
Use these to collect enamel residue when you wash it. The coffee filter and jar setup (see page 29) makes it easy to save washed enamel.

• Cotton, linen, or lint-free brown paper towels
Use high-quality, lint-free paper or cloth products when enameling so the enamels remain clean.

• Scouring powder or pumice
This inexpensive supply makes cleaning easier, but is not necessary.

• Green kitchen scrub pad
Green kitchen scrub pads can be very helpful when working with metal. Always make sure the pad you're using is clean. Never use steel wool because it easily contaminates enamel.

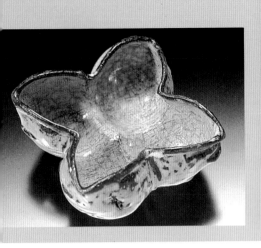

Left: Joan MacKarell *Rock Necklace*, 2004. 1⅕ x ⅘ in. (3 x 2 cm) each bead; 17⁷⁄₁₀ in. (45 cm) total length. Copper, enamel, jasper beads. Photo by James Austen.

Far left: Maria Phillips *Within*, 2002. 2½ x 2 x 1 in. (6.4 x 5 x 2.5 cm). Copper, silver, enamel. Photo by Doug Yaple.

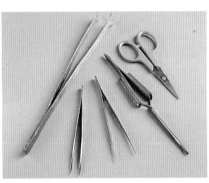

• **Metalworking tools** (above and left) For forming metal or setting enamels, you may want to have a basic selection of metalworking tools available, such as hammers, a jeweler's saw frame and blades, stakes, files, and drill bits, to name a few.

• **Scissors**
A small and sharp pair of scissors will come in handy for cutting metal foils.

• **Tweezers**
It's good to have several different types of tweezers; just make sure to keep one pair pointed and clean for cloisonné work.

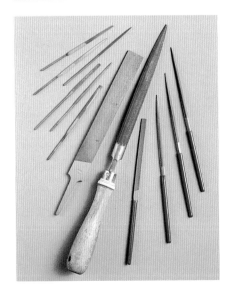

• **Sifters**
You can buy commercial sifters for powdered enamels or make your own using mesh screens and plastic or metal containers.

• **Watercolor paint tray or a selection of plastic spoons**
You need a clean, smooth spoon or tray in which to place and mix

your painting enamels. Choose a paint tray with a tight-fitting cover if possible.

• Sable paintbrushes and other assorted brushes

I recommend a 000 sable for painting techniques and a ½-inch (1.3 cm) soft flat sable brush for applying holding agents. Change the brush size to suit your needs.

• A spatula or small tool with a flattened or domed end

These tools are specifically shaped for picking up small amounts of enamel and can be purchased or made.

• Holding agents

These materials help adhere enamel powders to metal before firing. Water-based holding agents are also known as gum binder or gum tragacanth. Oil-based agents include oil of lavender, thinning oil, and squeegee oil.

• Ball Clay

This substance, also known as Scalex, can be painted on copper surfaces to prevent the formation of firescale.

KILNS

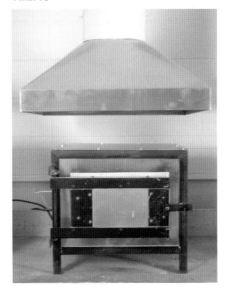

An electric kiln is a furnace structure made of firebrick and heat conducting coils of nichrome wire. The door, floor, ceiling, and walls of the kiln are well-insulated so the kiln will maintain an even distribution of heat throughout the firing chamber. The size of the kiln you need depends on the type of enameling work you plan to do. If you have the space, I recommend getting the largest kiln you can. This gives you the flexibility to create any size work. A tall and narrow kiln chamber is important for enameling goblets and vessel forms. A low horizontal kiln chamber works well for shallow vessels or trays. Small kilns are fine for enameling jewelry-scale pieces, and they can operate at 110 volts. Kilns with larger chambers usually

run on 220 volts of electricity. You can even use old furnaces for enameling as long as they will hold the heat at 1500° F (815° C).

The kiln elements (the coils of nichrome wire) can burn out if you heat the kiln too hot or if the elements are not evenly threaded in the firebrick wall. Fortunately, you can fairly easily replace the elements (see photo above). Depending on the age of the furnace, however, you may need to have new elements custom-made by an electrician. You can also use a kiln that has its elements covered by a castable refractory material.

Temperature Indicators

A *pyrometer* is an instrument that indicates the temperature in the firing chamber of a kiln. It's helpful but not essential. *Digital control pyrometers* hold the kiln at the correct temperature, eliminating the need to make manual adjustments as the interior temperature rises. If your kiln doesn't come with a pyrometer, you can purchase one separately and place it on top of or near the kiln. The *thermocouple* is the wire element that extends from the pyrometer into the kiln, relaying temperature information back to the pyrometer. Thermocouples deteriorate over time and may need replacing. If your kiln seems to be running hotter than its pyrometer indicates, a faulty thermocouple may be the culprit.

Kiln Doors

You may have a preference as to how a kiln door opens, so keep this in mind when making your selection. I prefer a door that opens to the side, but I've also used downward opening kiln doors with success.

Kiln Maintenance

Enamel that spills off pieces can build up on the floor of a kiln. This is common on old furnaces, especially ones used by many people. This buildup can make removing firing screens from the furnace difficult; they seem to "stick" to the glass spilled on the furnace floor.

Placing a kiln shelf on the floor will help protect its soft firebrick surface. To repair the floor or shelf, use a spatula or a butter knife and carefully scrape and flake off the glass as shown above. Try not to create craters and holes in a firebrick floor. If you happen to make gouges in the floor, use kiln cement to repair them. Mix the cement with water to a paste consistency. Use a wet sponge to saturate the kiln floor with water. Use a spatula, palette knife, or trowel to apply the kiln cement to the holes and uneven areas on the kiln floor. Allow the cement to dry for 24 hours, and then apply kiln wash to the floor to protect it during future spills.

I recommend coating all enameling furnace floors with kiln wash, even new ones or kiln shelves. Kiln wash is a mixture of kaolin and flint. You can purchase it from a ceramic or enamel supplier or perhaps get some from a potter friend or a ceramics department. (Potters coat their kiln shelves with kiln wash to prevent pieces from sticking during glaze firing.)

Applying Kiln Wash

1. Sweep out or vacuum debris from the kiln.

2. Mix the kiln wash with water until it has a creamy consistency.

3. Using a wide brush, paint the kiln wash onto the kiln floor. Use long

strokes and paint only in one direction, from back to front as shown. Let the kiln wash completely dry.

4. Using long brush strokes, paint another coat of kiln wash onto the floor. This time paint in the opposite direction, from side to side. Let the kiln wash completely dry.

5. Repeat steps 3 and 4 from four to five times until you build up a good layer of kiln wash on the kiln floor.

Note: The kiln wash may easily flake, but that's a good thing. When enamel spills cover the kiln floor, you can simply remove them with a spatula without digging into the firebrick and leaving holes.

TRINKET KILNS

You can use a trinket kiln for very small enamels, though you may find them somewhat limiting. You'll need to use a mica backing or a homemade firing screen when enameling in a trinket kiln. Its chamber is too small to accommodate most common trivets and screens.

TORCHES

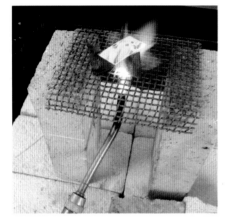

For soldering and for torch-firing enamels, I use an acetylene torch that combines with the oxygen in the air when lit. A propane torch is also suitable for torch-firing enamels. I encourage you to experiment with whatever type of torch setup you may already have, and try various size tips to fuse the enamel as quickly as possible.

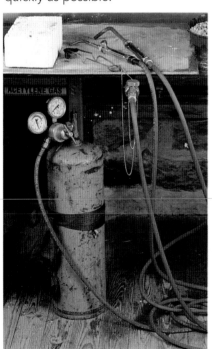

Above: Brooke Marks *Pendants*, 2003. 3½ x ½ x ½ in. (8.9 x 1.3 x 1.3 cm). Copper, enamel, leather. Photo by artist.

Right: Maria Phillips *Spire*, 2002. 32 x 4 x 4 in. (81.3 x 10.2 x 10.2 cm). Steel, silver, enamel. Photo by Doug Yaple.

SAFETY MATTERS

• *When sifting any type of enamel, always wear a mask that is approved for protection from fine toxic dusts. Inhaling enamel dust can be irritating to the nose, throat, and lungs. Many types of enamel, even lead-free colors, contain toxic ingredients such as cobalt, chrome cadmium, and selenium, which are all known carcinogens.*

• *Fire enamels in a well-ventilated area, preferably with a good exhaust system located near or over the kiln or behind the torch (see photos, right). Metal oxide fumes can be dangerous, and you should always be concerned about proper ventilation. This is especially important when you torch-fire because you're more likely to breathe the fumes.*

• *Never eat, drink, or smoke in the enameling studio.*

• *Frequently damp-wipe the surfaces in your enameling studio, and vacuum or damp-mop. This protects your work from contaminants and keeps you from inhaling or ingesting toxic dusts that settle on work surfaces.*

• *Personal hygiene is very important. Frequently wash your hands while you work. Shower and launder your clothes after each workday.*

Base Metals for Enameling

You can enamel on many different metals, but the most frequently used are copper, fine silver, sterling silver, gold, gold alloys, iron, and steel. When heated and cooled, both metal and enamel expand and contract. If the enamel's coefficient of expansion is not lower than that of the metal, the enamel can crack or flake off the metal. The melting point of the metal must always be higher than the firing temperature of the enamel, or the metal will melt when the glass is fused. When selecting metal to enamel, be sure you know its composition. Some metal alloys contain zinc, which can make the enamel pit or discolor with frequent firings. If you're in doubt about how a metal will react with enamel, do a test! I encourage you to experiment with different metals. The following information will help you in making your selection.

COPPER

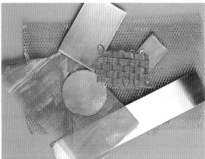

Copper is the most popular metal to enamel because it's inexpensive, malleable, can easily be cut and formed, and responds well to a wide range of enamels. (See gallery photos, left on this page and right on the facing page.) Copper is excellent when using opaque colors. It's also beautiful when clear enamel or some other transparent colors are fired directly onto its surface. When copper is heated, a brownish-red oxide known as *firescale* coats its surface in areas not covered with enamel. You must take this into consideration when you design a piece. You can remove firescale between firings, or you can incorporate areas of firescale into your design. Both of these options are discussed further on pages 42 and 43.

FINE SILVER & STERLING SILVER

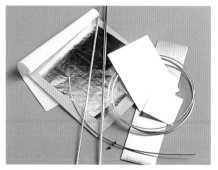

Fine silver is a good choice for enameling because the white color of the metal is brilliant and reflective when seen through transparent enamels (see gallery photo above, right). Because the metal is 99.5 percent pure silver, its surface does not become coated with any oxide or firescale when heated. This characteristic simplifies cleaning between firings and enhances the brilliance of transparent enamel colors. The downside of using fine silver is that it's very soft, and it warps easily when heated and cooled. One way to counter the stress of warping and protect an enamel surface from cracking is to form the metal to give it structural strength. Another way to prevent fine silver from warping is to provide it with

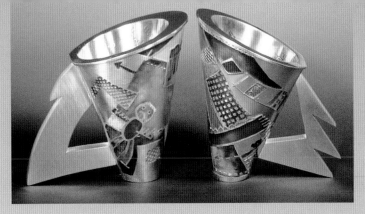
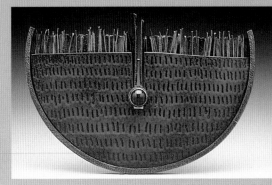

adequate support when it's fired. To achieve similar color effects, you can layer thin fine silver foils on top of other base metals. These metal foils are sold in booklets from enamel supply companies. (Refer to page 80 for further information on working with foil.)

Sterling silver, an alloy of 925 parts fine silver and 75 parts copper, is easy to enamel. For making vessels and larger objects, sterling silver is sometimes more suitable than fine silver because it warps less when heated and cooled (see gallery photo above, left). Though transparent enamel colors are slightly less brilliant on sterling silver than they are on fine silver, they are still quite beautiful. Because sterling silver contains copper, firescale appears when the metal is heated. Prior to enameling, you'll need to prepare the surface of any sterling silver piece to create a layer of fine silver. (Step-by-step instructions for this technique, known as *depletion gilding*, are on page 26.)

24-KARAT GOLD & 18-KARAT GREEN GOLD

Enameled gold is spectacular. Pure gold, or 24-karat, is the most brilliant of all metals under transparent colors (see photo page 24). Transparent warm colors are especially beautiful over gold because the yellow surface of the metal reflects through the glass. Because 24-karat gold is very soft and very costly, it's not often used in contemporary enameling. Alloys that consist only of gold, fine silver, and pure copper are more easily enameled than pure gold. I prefer enameling on 18-karat green gold, also known as non-tarnishing gold. It's alloyed with 25 percent fine silver. To achieve color effects similar to enameling on gold at a fraction of the cost, you can layer transparent colors over 22-karat to 24-karat gold foils. Gold foils are available in booklets from enamel supply companies.

LOW CARBON STEEL & IRON

You can purchase low carbon steel and iron, pre-coated with a base coat of white enamel over a dark ground, or with only a dark ground coat. Pre-coated steel plates are easy to use with the same enamels that fire well on copper, silver, or gold. Because it retains its shape with little warping when heated, steel is particularly appropriate for large, flat panels, murals, or architectural enamels. You can purchase special enamels that are suitable for

working directly on uncoated steel or iron if they contain less than .02 to .04 percent carbon (see photo page 24).

STAINLESS STEEL

Grade 304 stainless steel requires the use of high expansion enamel. It's possible to enamel on grade 410 stainless steel with low expansion enamel. Unlike other steels and irons, stainless steels do not require the application of a ground coat. If you want to work on this very rigid metal, ask your supplier about enamels specifically manufactured for this purpose.

BRONZE, BRASS & NICKEL SILVER

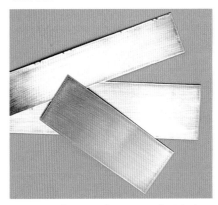

Because bronze, brass, and nickel silver are alloys that usually contain zinc, they can be difficult to enamel. With minimal firing, however, you may have some success. Multiple firings can cause pits or bubbles to appear in the enamel.

GILDERS METAL

Gilders metal is an alloy that commercial badge makers often use. It contains 95 percent copper and only 5 percent zinc. Most of the enamels you can use on copper or silver also work well on this alloy.

PLATINUM

With some experimentation, you can enamel on platinum. It's not one of the most frequently used metals because of its high cost. Some artists have found that lead-free enamels work better on platinum.

ALUMINUM

You can easily enamel some aluminum alloys with special low temperature, high-expansion enamels. (These aren't the same enamels used on copper, silver, and gold.) In addition to the special enamels, there are specific techniques you can use to achieve the greatest success, such as sifting on hot aluminum. If you're interested in enameling on aluminum, inquire about its application and firing procedures when you order the special enamels.

TITANIUM

Because of titanium's low expansion rate, you'll need to experiment to determine which types of enamels will work well. Opaque enamels are more suitable to use on titanium. Transparent colors will turn gray from the oxides in the metal.

SELECTING THE METAL

Knowing how an enameled piece will function, how large it will be, and how it will be formed helps you to select the correct type and *gauge*, or thickness, of metal. Whether you heat it in a kiln or with a torch, metal expands and contracts. Enamel also expands and contracts when it's heated and fused to metal, but always to a different degree. To test this, fire a coat of enamel on one side of a long narrow strip of flat metal, and watch it curl up (see photo above). Even when you counteract this by applying an enamel coat to the back side of the metal, there will still be some warping. As the metal cools it contracts more than the enamel, which becomes rigid below its softening point. The metal, therefore, begins to "pull up."

Metal	Melting Point (Farenheit)	Melting Point (Celsius)
18-karat green gold	1770°	966°
24-karat gold	1945°	1063°
Aluminum	1220°	660°
Brass	1749°	954°
Bronze	1945°	1060°
Copper	1981°	1083°
Gilders metal	1950°	1065°
Iron	2793°	1535°
Low carbon steel	2750°	1511°
Nickel silver	2020°	1110°
Platinum	3225°	1774°
Silver, fine	1762°	961°
Silver, sterling	1640°	920°
Stainless steel	2500°	1371°
Titanium	3272°	1800°

Solder Type	Flow Point	Contents
Eutectic Solder	1460° F (793° C)	71.9 percent fine silver 28.1 percent copper
IT Solder	1490° F (809° C)	80 percent silver 16 percent copper 4 percent zinc
Hard Solder	1425° F (773° C)	76 percent silver 21 percent copper 3 percent zinc
Medium Solder	1390° F (747° C)	70 percent silver 20 percent copper 10 percent zinc
Easy Solder	1325° F (711° C)	60 percent silver 25 percent copper 15 percent zinc

HOT TIPS
METAL THICKNESS & WARPAGE

— Before choosing the gauge of the base metal, consider the technique you'll be using and the thickness of enamel you plan to apply. If the final enamel coat will be thin, you can usually use thinner gauge metal.

— In cloisonné enameling (see page 105), you'll build up many layers of enamel in order to reach the height of the cloisonné wires you use. Unless you form the metal, you might need to use 18-gauge metal for small cloisonné jewelry pieces. If you form the metal, you can use 20- or 24-gauge. When you etch metal to create recesses for the champlevé technique (see page 114), you'll need to select fairly heavy gauge sheet metal. The added thickness provides plenty of room for you to deeply etch the metal while continuing to afford sufficient structural strength.

— Thicker metal "pulls up" less than thinner metal. If you structurally strengthen metal through forming, it will pull up even less.

— Since flat metal warps more than formed metal, prior to enameling you may want to form pieces to create greater structural strength. This preparation will allow you to use thinner gauges of metal, which is often desirable to lessen the weight of jewelry. When forming a piece of metal by dapping in a wood or

steel block (see photo A), using a hydraulic die press, fold forming, or otherwise shaping it with a hammer, I use 20-, 22- or even 36-gauge stock, depending on the size or the structure of the finished form. If you want the metal to remain flat, use 18 gauge for small 2- to 3-inch (5 to 7.6 cm) sample pieces, and up to 16- or 14-gauge metal for larger wall pieces. (For really large pieces you may want to work on steel, which warps less than most other metals.)

— Since an alloyed metal will expand and contract to a lesser degree than metal in its pure state, this fact should be taken into consideration when selecting which metal to use for a project. For example, I have found that on large, fine silver vessels the enamel cracks more often and is less stable than on sterling vessels (sterling silver, is an alloy of fine silver and copper, and is more rigid and less likely to move dramatically when heated and cooled). I usually choose fine silver for small cloisonné jewelry pieces because it does not

firescale when heated and the enamel colors look more brilliant.

— Enameling adds weight to metal, so carefully consider how the piece will function before selecting the correct metal gauge. For example, wearing large, heavy 16-gauge metal earrings coated with enamel in pierced ears would be difficult!

— Use thin enamel coats on thin gauge metal. Thick coats of enamel are more likely to crack or chip off the surface.

— You can help prevent warping on any gauge metal by properly supporting a metal piece during firing. Place the support points of a firing trivet directly under the areas where warping is most likely to occur (see photo B).

— Once you understand these principles, you'll be able to enamel on wire, sheet, properly prepared castings, precious metal clay, mesh screen, woven strips and screen, copper pipe, small and large tubing, and much more. Research your options, and make tests!

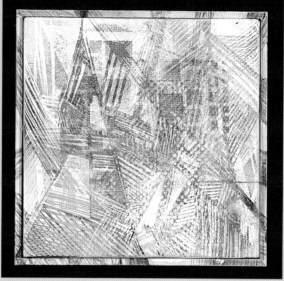

Left: Jo Ann Tanzer *Suspended From a Fixed Point*, 2001. 14 x 14 x 2 in. (35.6 x 35.6 x 5 cm). Steel, enamel. Photo by Kimberly Williams.

Far left: John Paul Miller *Moth with Tails*, 3 x 2¾ in. (7.6 x 6.9 cm.). 24-karat gold, 18-karat gold, enamel. Photo by artist.

Solders to Use with Enameling

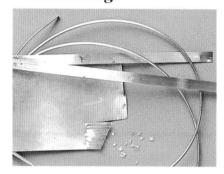

When using solder to join a metal piece before enameling, you should pick a solder that melts at a higher temperature than the enamel fuses. If you don't, the piece will fall apart in the kiln. If you apply enamel over a solder that contains zinc, it may cause the glass to pit, discolor, or flake off. You can successfully use solders on enamel pieces if you consider their melting point, what they are made of, and when you plan to apply the solder (prior to, during, or after firing).

If you plan to enamel over a solder seam, be sure to use eutectic solder, an alloy of pure silver and copper that contains no zinc. You can fire colors directly over a eutectic solder seam or joint without the enamel pitting. If you're working on sterling silver, heat and pickle the piece to bring the fine silver in the solder to the surface. (This process of depletion gilding is fully described on pages 26 and 27.) By preparing the soldered metal this way, you'll be able to apply transparent colors directly over the seam without it showing through. If you're working with opaque colors, however, depletion gilding the solder seam isn't necessary.

Prior to enameling, if you solder a piece in such a way that the enamel will not touch the seam, you can use eutectic solder, IT solder, or hard solder.

Use medium or easy solder on pieces you want to solder in the kiln while enameling.

If you're torch soldering a piece after enameling, depending on the setup and placement of the joint, I often use hard solder. The high heat required for the hard solder will re-flow the enamel as if it were being re-fired in a kiln.

You can also use easy solder on a torch-soldered piece if the joint is far away from the enameled surface. Sometimes when you heat a piece to the melting point of easy or medium solder, the enameled surface will re-fire to the orange-peel stage or possibly crack. If possible, and the construction of the piece allows it, I prefer to use hard solder and completely re-flow the glass to its fusing temperature.

For futher information on how to solder with enameling techniques, see page 55.

Cleaning & Preparing Metal

Whichever enameling technique or metal you choose, there's one important rule to follow: the metal surface should be completely clean before any enamel is applied. Once the object you want to enamel is cut and formed, clean it immediately before applying the enamel.

There are several methods for preparing and cleaning the most frequently enameled metals. You can choose the appropriate technique based on the piece you're creating or the facility in which you're working. Use the guidelines I'm providing as a reference for researching and testing these methods.

Mark Hartung *Eight Enamel Brooches*, 1999. 1¼ to 3 in. (3.2 to 7.6 cm). Sterling silver, enamel. Photo by Robert Muller.

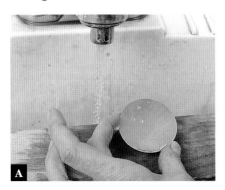

As the final step before enameling, always hold the metal under running water and check to see that the water doesn't pull away from any part of the metal surface. This test reveals any metal areas that aren't clean or grease free. If the water beads up or pools (photo A), clean the metal again until water sheets across its surface, indicating it's ready to enamel (photo B).

CLEANING COPPER

Here are six ways to clean copper.

• Sand the metal with emery paper (photo C) or a clean green kitchen scrub pad (photo D) under running water. Avoid steel wool as particles may adhere to the metal and cause pitting in the enamel.

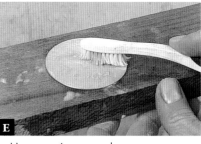

• Use pumice powder or powdered detergent with a toothbrush (photo E), scrub pad, or sponge to clean the metal. Make sure the detergent doesn't contain hand softeners (grease!) and that the scrub pad is clean. Rinse the metal well under running water.

• If you work in a well-ventilated metal studio, you can use a pickle solution of sodium bisulphate (commonly used as a swimming pool cleaner) or diluted sulphuric acid. (Dilute the solution at a ration of approximately four parts water to one part acid.) Heat the copper

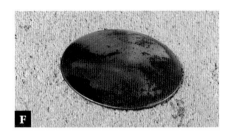

with a torch or in the kiln until a green shadow covers the metal and all grease burns off the surface (photo F). Put the metal into the pickle for a few minutes until clean (photo G), and then rinse it well under running water.

• Dip the copper in a dilute solution of nitric acid (one part acid to three to six parts water) for a few minutes until clean. (You shouldn't use the nitric acid solution on pieces with solder joints, however, because it will weaken those connections.) Thoroughly rinse the metal in baking soda and water to neutralize any acid that may remain in crevices, and then follow with a final rinse in plain water.

• If the copper isn't covered with too much firescale or grease, you can sprinkle regular table salt on its surface, wet it with vinegar, and then polish it with a clean kitchen scrub pad. Alternately, you can soak the copper in a solution of one part table salt to eight parts vinegar.

• Use a commercial copper cleaner. You might prefer these products to all other methods.

CLEANING FINE SILVER OR PURE GOLD

Here are three ways to clean fine silver or pure gold.

• Heat the metal with a torch or kiln to remove grease, and let cool. Thoroughly rinse the metal under running water.

• Use emery paper or an abrasive compound such as pumice to remove oil or grease. If the metal loses its shine from the abrasive action, shine it again with a soft brass or glass brush prior to enameling.

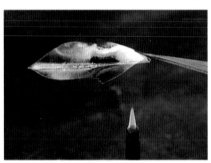

• To make pure metal very shiny, you can heat it with a torch until the surface liquefies or "flashes" brightly. During this process, the top layer of the metal essentially melts. This brilliance is spectacular under transparent enamel colors.

CLEANING STERLING SILVER

To prepare sterling silver, I recommend using the depletion-gilding technique to bring the fine silver to the surface. Sterling silver contains copper oxides. By heating it, you bring these oxides to the surface. You then put the heated metal in a pickle solution of sodium bisulphate or dilute sulfuric acid to clean off the firescale. After pickling, gently burnish the metal with a very soft brass brush and liquid soap. Repeat this process several times until no firescale appears and a thin skin of fine silver forms on the metal when heated. If you like, burnish the metal once again with the soft brass brush, prior to enameling.

You can use a torch or a kiln to bring firescale to the surface of sterling silver. I usually use a torch because I like to watch the piece while I'm heating it, using special care not to re-flow solder joints. You can also place sterling silver into a low temperature kiln, approximately 1200° F (649° C), for up to five minutes or until gray firescale forms on the surface of the metal.

DEPLETION GILDING

1. Use a torch or a kiln to heat the sterling silver until copper oxides come to the surface. (Resist the temptation to overheat the metal. The oxides may only appear as a slight gray shadow. Heat gently, take the torch away, heat gently, and take the torch away again. You'll only see the oxide shadow when the torch is removed from the piece.) If you heat the sterling silver in a 1200° F (649° C) kiln, it may take from two to five minutes to bring the firescale to the surface.

2. Pickle the metal until it appears white and frosty.

3. Under running water, use a very soft brass brush lubricated with a drop of liquid soap to burnish the surface of the metal. Dry the metal.

4. Repeat steps 1 through 3 as many times as necessary to deplete the copper oxide from the sterling silver surface, leaving a fine silver layer. When the silver remains frosty white when heated, you'll know you have achieved a sufficient layer of fine silver.

5. If you want to create more shine, you can carefully burnish the finished sterling silver surface with a soft brass brush or glass brush before enameling. Work gently; using the soft brush with too much force can abrade the fine silver off the surface.

CLEANING GOLD ALLOYS

Some gold alloys produce firescale or turn dark when heated. To prepare the surface for enamel, you can use a technique similar to depletion gilding silver. Heat the gold, and then pickle it in a solution of two or three parts water to one part nitric acid or use a sodium bisulphate solution. Repeat the heating and pickling process until the gold shows no discoloration when heated. Rinse well.

FAST FACTS & HELPFUL HINTS

— Anytime you use a pickle or an acid bath to clean metal you may want to alkalize its surface by dipping it in a baking soda and water rinse or by using saliva. Rinse the alkalized metal in clean water.

— If you plan to use transparent enamel directly on metal, be certain its surface appears exactly the way you want it to look beneath the color. If you want to see shine beneath the color, be sure the metal is bright!

HOT TIPS
DEPLETION GUILDING

— At some point during the depletion gilding process, firescale will no longer appear on the surface of the sterling silver. Take care not to heat the silver too much while you're mistakenly waiting for the firescale to appear. If you polish the metal surface with the brass brush between heatings as directed, you'll receive a visual indicator of the proper heating temperature: the metal surface will change from a shiny to a matte finish. This is subtle, but you can actually see the shine leave the surface.

— You may be able to create another visual indicator, depending on how a particular piece is

constructed. File or sand a small place on the sterling silver in an area that won't be enameled (photo C). Heat the metal only until the abraded area becomes dark with firescale (photo D). When the firescale appears in that spot, you'll know the rest of the metal was brought up to the proper heating temperature; you can stop heating the metal and put it in the pickle.

— To ensure a good coating of fine silver, I recommend heating a sterling silver piece about three additional times after the firescale disappears from its surface. The entire heating and pickling process can take anywhere from three to 10 times, depending on the metal techniques you used to create the piece.

Left: Jane Short *Weavers Bowl: The Worshipful Company of Weavers,* 2002. 15 in. diameter (38.1 cm). Silver, enamel; champlevé, basse taille. Photo by Clarissa Bruce.

Far left: Valeri Timofeev *Box,* 2001. 1½ x 3 x 2¼ in. (3.8 x 7.6 x 5.7 cm). 18-karat gold, diamond, enamel; plique-à-jour. Courtesy of a private collection. Photo by artist.

Preparing Enamel

To determine how to prepare the enamel before applying it, carefully study the object you're making. Is it flat or three-dimensional? Will a holding agent be needed to help the enamel adhere to the metal surface? Since smaller particles create less bounce when sifted and adhere better to metal edges, would it be better to sift with 80-, 150- or 200-mesh enamel? Would it be easier to spray, paint, or pour on liquid enamel? Once you thoughtfully study the piece, you can prepare the enamel.

POWDERED ENAMELS & PARTICLE SIZE

Enamels are manufactured in various grits, or particle sizes. The grit size is measured by passing the powdered glass through a mesh screen with a standardized number of openings in 1 linear inch (2.5 linear cm). The more openings in a screen per inch (or centimeter), the smaller grit it will filter. You can make or purchase your own 60-, 80-, 100-, 200-, and 325-mesh screens. With them you can sift and separate enamel powder, typically purchased in the 80-mesh grit size, into smaller particle sizes for different purposes.

If you want to use a variety of enameling techniques with ease, it's important to understand particle size. For applications that require great enamel clarity, such as wet packing over gold and silver, working with a larger particle size, such as 80 mesh is advantageous because the fired glass will be more transparent. For applications such as fine painting, you may prefer using a smaller particle size, such as 200 or 325. Finer enamels are less grainy and flow more smoothly. Prior to sifting, if you remove enamel particles larger than 150 mesh, the enamel is less likely to bounce off the edges of the metal. This preparation can even eliminate the need to use holding agents to adhere the sifted enamel to the metal before firing. (For more information on holding agents, see page 31). However, if the particle size of a sifted base coat is too fine, it will likely pull away from the edges and toward the center of a metal piece.

SEPARATING MESH SIZES

The first step in preparing enamel is to separate the correct particle size for your application. You can use a sequence of screens to sift particles to the desired size.

1. Stack the enamel sifters of various mesh sizes. Arrange them so the sifter with the largest mesh openings is on top, and the finest mesh is on the bottom, directly above a collecting tray.

2. Place a quantity of 80-mesh enamel in the top tray.

3. To keep dust down, cover the top sifting tray with another tray or lid.

4. Shake the stacked sifters. The enamel will fall through the screens, according to its particle size. The largest particles will remain in the top tray. They will be the most transparent enamel and should need little or no washing, depending on their use. The finest enamels fall to the bottom collecting pan. (Save these for painting projects.)

WASHING ENAMELS

Commercial enamel manufacturers use ball mills to grind their stock. By-products of this process include porcelain dust, a fine enamel dust residue. These impurities affect the clarity of transparent enamels, causing them to appear cloudy and less brilliant. Even if you remove the fine powders by pre-sifting commercial enamel through 200- or 325-mesh screens, you'll need to wash the remaining enamel for

Above: Sarah Turner Untitled, 2002. 5 x 6 in. (12.7 x 15.2 cm). Copper, enamel. Photo by artist.

Right: Nancy Bonnema *Shuttle*, 2002. 4 x 1 in. (10.2 x 2.5 cm). Sterling silver, enamel, riso screen. Photo by Doug Yaple.

optimum clarity. (This is especially important when luminous transparency is critical to the work, such as firing transparent colors over gold foils, silver foils, or textured surfaces.) If you want to sift out the fine enamel to save for painting projects or counter enameling, do this before washing it. Because washed enamel has a shorter shelf life, wash only the amount you need relative to the size of a piece. I usually wash two to three ounces (56.6 to 85 g) at a time.

How to Wash Enamel

1. Place the enamel color you're washing in a clear container (glass or plastic), and add at least twice the amount of water. Cover the container with a tight-fitting lid, and then shake or swirl it so the water becomes cloudy.

2. Allow the enamel to settle on the bottom of the container until a fine line settles on the top of the enamel. Pour off the cloudy water into a second container. (If you want to save the fine particles as you pour off the cloudy water, use a coffee filter as a strainer.)

3. Repeat steps 1 and 2 until the water you pour off is clear. I recommend using distilled water for the final two washings in case the tap water you're using contains chemicals that might cause the enamel to break down more quickly.

HAND-GRINDING ENAMEL

Some enamelists grind and wash enamels from lumps each day. This procedure assures that their colors are absolutely clear and bright. To grind your own colors, put about 1 ounce (28.3 g) of lump enamel in a mortar. Use a pestle to pound the glass down for 30 seconds as shown, crushing the lumps. (It may be helpful if you start by putting the particles in a plastic bag and pounding them with a hammer.) Cover the pounded particles in the mortar with a cloth, or put them in a plastic bag that has a hole in it through which the pestle can pass. Pound with the pestle using a rocking, crushing motion, not a circular grinding motion. After 30 seconds sift the particles through an 80-mesh screen. Place half of what is left on top of the screen in the mortar, and put the other half aside. Grind the glass for about 30 seconds, this time using a rotating motion. Sift the enamel back through the 80-mesh screen, and put what is left on the screen back in the mortar. Add half of what was put aside back to the mixture. Repeat grinding and sifting until complete. The reason for removing and sifting the enamel so frequently is that the larger particles help to grind the smaller particles. If you did not remove the smaller particles dur-

ing the procedure, they would be "over ground," becoming smaller and smaller. You can add a very little water or use a cloth cover to keep the dust down.

STORING PREPARED ENAMELS

Washed enamel does not have a long a shelf life, especially when stored underwater. You can store wet enamels in airtight containers for short periods of time if you cover them in distilled water. Never allow enamels to slowly dry out or sit in a state of dampness. This will change their composition.

The photo above shows white specks on an enamel that has been stored in water for several months.

Keep in mind that all ground enamels, even those stored dry, will absorb moisture from the atmosphere and can decompose over time, and some decompose more readily than others. Opaque enamels seem to decompose sooner than transparents and this decomposition will vary depending on the manufacture of the different enamels. The following information is helpful to know as you store and save your enamels:

• Lump enamel will decompose less quickly than powdered enamel.

• The longer an enamel is exposed to air and water, the more quickly it will decompose.

• If dry enamels are kept in airtight containers, they will last a long time.

• Washing or regrinding decomposed enamel with water or acid may or may not remove decomposition.

DRYING WET ENAMELS

1. Place a coffee filter over a clean and empty jar. Shake the container of recently washed enamel, and quickly pour its contents through the coffee filter. The water will drain through the filter, and the enamel will settle at the bottom.

2. Pick up the filter and place it on top of a sheet of aluminum foil. Fold up the sides of the foil to make a small tray.

3. Position another sheet of foil over the tray to keep the enamel clean. Place the washed enamel in the tin foil tray on top of a hot kiln to dry.

FAST FACTS & HELPFUL HINTS

—Wash several colors at once. While enamel is settling in one container, you can start shaking or adding water to another container.

—Opaque colors don't need washing unless they're very dark and appear cloudy when fired.

—Transparent colors fired directly on copper don't need washing. When fired at hot temperatures, the colors become clear quite easily.

—Always wash enamels you plan to layer over other colors as well as enamels you want to fire on silver or gold.

Katy Bergman Cassell *Hollow Ball Necklace: White*, 2002. 18 x 1¾ in. (45.7 x 4.4 cm). Copper, sterling silver, enamel. Photo by artist.

Sarah Krisher *Formal Arrangement*, 2002. 2 in. (5 cm), 3½ in. (8.9 cm). Copper, silver, enamel. Photo by Robert Muller.

Applying Enamel

THE IMPORTANCE OF AN EVEN BASE COAT & COUNTER ENAMEL

Coating both sides of the metal with enamel evenly distributes the stress of expansion and contraction, can prevent cracking, and can lessen warping. The coat on the front of the piece is called the base coat, and the coat on the back is the counter enamel. For most techniques, it's important for the base coat to be as even and complete as possible. (An exception to this would be when you leave copper areas exposed to intentionally create firescale.) It will warp less if the counter enamel is about as thick as the base coat and has the same coefficient of expansion. Before applying the counter enamel or the base coat, decide whether or not to use a holding agent to better adhere the enamel to the metal during application and firing.

USING HOLDING AGENTS

A holding agent is an organic binder that makes it easier to control and work with the enamel during application. Holding agents also help adhere the powdered glass on dimensional, sloped, or vertical surfaces during firing. Sometimes a holding agent isn't necessary, such as when you apply enamel to a flat surface, but it is important to use one when working on a wire structure, for example, or on the steep sides of a vessel. You should use holding agents sparingly and only when required. Before you fire a piece, make certain the binder is completely dry. It must also completely burn out so ash and gas bubbles aren't trapped in the enamel layers. (These could surface during later firings.) Although you gain greater control by using holding agents, you'll sacrifice some clarity with transparent colors.

Use water to dilute by half a water-soluble holding agent before using it to apply a base coat or counter enamel. Immediately before sifting,

A

B

apply the binder to the metal surface with a soft flat paintbrush (photo A) or a pressurized spray bottle (photo B). (There are other oil based holding agents such as squeegee oil or oil of lavender that are often used with painting enamels, but in most cases I do not use them to hold base coats and counter enamels in place.) When creating the complex dimensional forms pictured in the gallery photo above, Katy Bergman Cassell found that squeegee oil was the best binder to use for the sifted enamel because it dries slower than water-based binders.

SIFTING

It's very easy to evenly apply enamel using the sifting technique. You'll use specially prepared screens of various mesh sizes and sift dry enamel through them onto metal surfaces.

Preparing to Sift

Before you begin, prepare a clean area for sifting, wear a dust mask for safety, and have plenty of clean magazine pages nearby to catch excess enamel. By using slick magazine paper rather than newsprint, you can easily pour extra enamel back into its container. If you place a second clean page beneath the one on which you're sifting, it will keep the back side of your catch paper clean. Otherwise, when you make a pouring funnel to return the

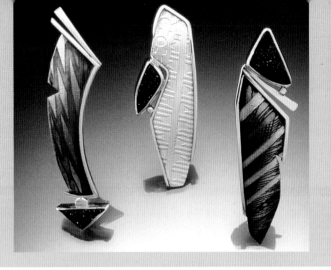

enamel to its container, stray dirt or colored enamel particles could fall from the back side of the catch paper and contaminate the enamel. Protected by the second page, the back of the catch paper stays clean!

Sifting a Base Coat or Counter Enamel

1. Choose a sifter that's the appropriate size for the piece you're enameling. You wouldn't want to use a very small sifter to sift enamel on a very large metal piece or a very large sifter on a tiny metal piece.

2. Fill the sifter no more than one-third full of dry enamel. This makes the particles fall through the screen more easily.

3. Spray or thinly paint a holding agent onto the piece if its shape requires one.

4. Sift the enamel onto the metal in an even rhythm, using overlapping paths. The movement of the sifter makes the enamel particles fall through the screen. To accomplish this I hold the sifter against my third, fourth, and fifth fingers. I rest my thumb on the top of the sifter for control and my index finger against its side. I tap the sifter with my index finger, and depending on the size of the piece I'm sifting, I also might use my wrist or even my elbow to create more movement.

5. Begin at the outside edge of the metal and sift in concentric circles toward its center. Sift extra enamel on the outer edge, where, during hot firings, the enamel is more likely to pull away.

6. Always keep the sifter at a 90-degree angle to the worktable. Tilt the metal piece if necessary, not the sifter.

7. Hold the metal in your hand, resting it on the top of your fingers when you sift. This way, immediately following sifting, you can easily transport the piece to a trivet in preparation for firing. If the piece is flat on the table while sifting, the enamel on its edges may be disturbed when you pick it up to move it to a trivet. If you make it a habit to sift enamel while the piece is on the trivet, the trivet eventually becomes coated with enamel. It takes practice to learn to sift skillfully so the enamel coats the metal surface evenly, without excess.

Right: Sarah Perkins *Sophia Container,* 2000. 6 x 4 in. (15.2 x 10.2 cm). Silver, onyx, enamel. Photo by Tom Mills.

Far right: Veleta Vancza *Confiscation of the Square,* 2002. 16 x 16 x 3 in. (40.6 x 40.6 x 7.6 cm). Copper, wood, enamel. Photo by John Guillemin.

8. Re-apply a light spray of the holding agent after the first sifting if necessary. (I only recommend this for sifting on very dimensional pieces.)

9. Let the piece completely dry before firing if a holding agent was used. If you place the sifted enamel on top of a hot kiln or under a heat lamp, this takes only a few minutes.

Cleaning Workspace After Sifting

When you're finished sifting each color, fold the used magazine papers in half, and then in half again, and place them in the trash. Frequently wipe table surfaces. Keep the lids on containers whenever possible to prevent contamination by dust or other enamel colors. These studio practices help keep dust to a minimum.

WET INLAYING

When applying various colors next to each other for shading or special effects, using wet enamel gives you more control. It's also an easier way to apply enamel around soldered findings or in hard-to-reach places where other areas of the finished piece won't be enameled. I often use the wet inlay technique as a base coat or a counter enamel on dimensional or champlevé etched pieces.

Wet Inlaying a Base Coat or Counter Enamel

1. If the enamel you'll be using is dry, put a small quantity in a shallow dish or paint tray (photo A). Using an eyedropper, syringe, or squeeze bottle (photo B), add just enough water drops to cover the

dry enamel, and let the water soak into the glass. (Don't mix or stir the enamel. This creates air bubbles that could later be trapped in your piece.)

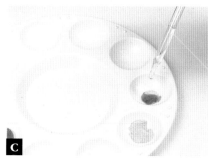

2. If the enamel you'll be using has been washed and stored wet, scoop a small amount of the wet color into a shallow dish or paint tray. Add a few drops of water-based holding agent to the wet enamel if necessary (photo C). Allow the binder to settle into the mixture. Do not stir it and create air bubbles.

3. Using a spatula and fine sable paintbrush (such as the 000 size) or a spatula and an inlay tool, pick up some wet enamel and hold it over the piece. Gently transfer the enamel mixture onto the metal and push it into place with the paintbrush or tool. (Although this is the traditional technique, I also inlay wet enamel without using the spatula. I simply use a paintbrush to pick up the color and directly transfer it onto the piece.)

4. If the enamel is too wet, it will spread out or flow down sloping sides. To remedy this, hold the corner or edge of a clean linen or lint-free cotton or paper towel next to the wet enamel. (Brown paper towels work well.) Let the towel absorb, or wick, the excess water. As you wet inlay, you may want to frequently wick the enamel, especially if you don't want adjacent colors to bleed.

5. As you wet inlay, gently tap the sides of the piece or the surface of the wet enamel to level it.

6. You also can smooth the surface with a bent wire tool or a paintbrush. Use a dabbing motion, not a painting or pulling one. It may feel like you're pushing around sand or sugar grains with the tip of the brush.

LIQUID ENAMEL
Dipping, pouring, painting, or spraying liquid water-base enamel is a similar technique to glazing pottery, and it works especially well for some pieces. Using liquid enamel makes it easier to coat the inside of hollow objects, such as containers or beads, because you can pour it in and out of forms. You can purchase this type of enamel in liquid form or as a dry powder that you mix with water before use. Some practice is necessary to apply an even coat of liquid enamel, but if it's the correct consistency and well mixed, you'll have an easier application. (Refer to pages 97–104 for ways to use liquid enamel to create special effects.)

Preparing & Applying Liquid Enamel as a Base Coat or Counter Enamel
Premixed liquid enamel is usually a good consistency for brushing or dipping. You can, however, add small amounts of water to the mixture until it reaches the consistency you like.

To prepare dry liquid-enamel powder, only use a small amount of water. Pour the powder into the water, as if mixing plaster (photo A). Let the powder soak for a few minutes, and then gently mix it with a spoon (photo B). Add more powder or more water to the mixture until it's the correct consistency for your application. Experiment to determine what works best for you.

You may need thinner liquid enamel if you're painting or air brushing it onto metal. Slightly thicker liquid enamel is often preferable for pouring techniques. I use the same finger-dipping test I learned for applying glazes to pottery. I dip my finger into the liquid, pull it out, and see how thickly the mixture coats my fingernail (photo C). For pouring the liquid onto dimensional surfaces or for dipping, I like the enamel to thickly coat my fingernail. The consistency should be thicker than heavy cream but not as thick as sour cream!

Hye-young Suh *Brooches,* 2002. 3 in. (7.6 cm) each. Copper, enamel; electroformed. Photo by artist.

Dipping Hold a clean metal piece with your fingers, with tongs, or with pliers. Gently dip it into a bucket or tray containing liquid enamel. Touch up any marks left on the piece before firing.

Pouring Use a cup, spoon, or a pitcher to pour the enamel onto the metal, rotating it until the liquid evenly coats the piece.

Painting For painting large, flat pieces of metal, use a soft, wide brush, such as the type house painters use for trim. To paint smaller pieces, use a flat sable artist's brush. Test and adjust the proportion of water to enamel as needed until you can evenly paint it onto the metal.

Spraying It's best to spray enamel in a well-ventilated booth, using compressed air and an airbrush with an orifice that's at least 1/20 inch (.14 cm) in diameter. You can also thin the liquid and spray it with an air-pressurized spray canister.

Storing Liquid Enamel

Liquid enamel dries out quickly so always keep it covered. It also can cake on the bottom of its container. If this happens, use a small hand mixer to restore the mixture to the correct consistency, or mix by hand with a spoon or spatula.

HOT TIP

—Sifting, wet inlaying, and liquid enamels are discussed here in reference to the fundamental practice of applying base coats and counter enamels. As you'll see in the Enameling Techniques section, within each of these methods there are many variations, and these techniques are often combined within a single piece. You can use wet inlay on pieces that have been sifted and stenciled. You can sift over pieces that have been wet inlayed or coated with liquid enamel. You can apply many enamel layers to a single piece, but in most cases, the layers following the base coat should be thin for greater control. Thick enamel is more likely to cause cracking and crazing. With multiple layers it may be necessary to counter enamel the back side of a piece more than once so both sides of the finished piece maintain the same coefficient of expansion. Doing this will make the enamel less likely to crack from warping.

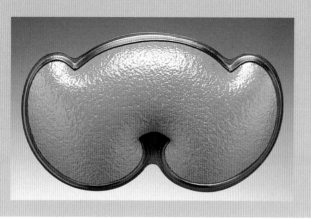

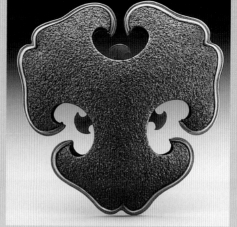

Left: Joe Wood *Ero-Blue #1*, 1995. 2¾ x 2¾ x ½ in. (6.9 x 6.9 x 1.3 cm). Copper, oxidized sterling silver, enamel. Photo by Dean Powell.

Far left: Joe Wood *Yellow Set #3*, 1995. 1¾ x 3 x ½ in. (4.4 x 7.6 x 1.3 cm). Copper, sterling silver, enamel. Photo by Dean Powell.

Firing Enamel

Once you apply the enamel to the metal, it's time to watch the glass fuse. This step involves fire, and it's exciting! The red-hot heat of the kiln or the dancing flame of the torch transforms powdered glass into beautiful transparent, translucent, or opaque colors. For me, watching the colors change and seeing the glass flow and fuse is one of the most exciting parts of the enameling process.

Though some enamellists will fire a piece while still damp with water (and no binder), I always play it safe and let the piece dry before firing. You can let it air-dry on a trivet, or to speed the drying, you can place it under a heat lamp or on top of a hot kiln. Prior to firing, the enamel on the surface of a piece should look powdery and dry. (With all the drying and cooling time required between steps, it's very efficient to work on several pieces at one time!)

TIMING

Learning to correctly fire enamel requires practice. There are no rules stating exactly how long each color or each piece should be heated, and therefore no timer can be set. You'll need to adjust firing times depending on several factors, such as: metal type, thickness, and size;

enamel color and thickness (remember, some colors have a different viscosity and softening point than others); the particle size of the enamel used (smaller particle sizes, such as 325 mesh, require less heat to soften, while larger particle sizes, such as 80 mesh, require more heat); the size and weight of the firing screen and trivet (screens and

HOT TIP
KILN FIRING TIME

— Think twice about comparing notes with fellow enamellists regarding firing temperatures and times. Even if you're using the same color on the same piece, it's unlikely that two kilns are exactly alike. Firing times can vary depending on the age of the kiln and its elements, the age of the thermocouple and pyrometer, and the location of the pyrometer in the kiln. The furnaces may pull different wattage or be different sizes. A lower wattage kiln will recover heat more slowly when the door is opened. Other variables include the size and weight of the firing trivets, the amount of insulation in the furnace, and the location of the piece in the kiln.

trivets act as a heat sink in the kiln—the larger and heavier the trivet or screen, the longer the enamel will take to fire); kiln or torch type; and piece placement in relation to the heat source (the back of the kiln may be hotter than the front).

FIRING SUPPORTS

If there is enamel on the back side of the piece you're firing, you can't place it directly on a firing screen, or it will fuse to the surface. Instead, choose a firing support that won't touch the enamel-covered surface. Also, try to be sure that this tool supports the metal in locations that help keep it from warping.

You can purchase trivets in different sizes and shapes. The photo below shows a variety of trivet shapes and

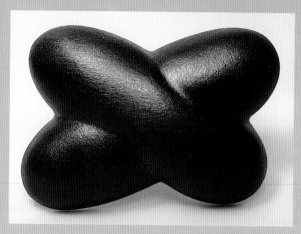

Right: David C. Freda *Study of Tomato Hornworm Caterpillar Brooch*, 2002. 4½ x 1¼ x 1¼ in. (11.4 x 3.2 x 3.2 cm). Fine silver, sterling silver, 24-karat gold, 18-karat gold, 14-karat gold, enamel. Photo by Robert Sanders.

Far right: Synnøve Korssjøen *Brooch*, 2000. 3 x 2½ x 1 in. (7.6 x 6.4 x 2.5 cm). Silver plate, enamel. Photo by Sigurd Bronger.

how different pieces can be supported. You can also buy special trivets for beads, but it's easy to make your own. Simply turn a screen upside down and thread a steel rod through its openings.

You can use sheet mica as a backing for some metal shapes that are difficult to support on a trivet. You can also use mica as a temporary backing for a plique-à-jour enamel as described on page 126.

WARPING

Although some metal warping is the result of non-uniform heating or cooling, it's most often caused by high temperature firing and improper support. If you give sufficient thought to determining the proper support for a piece prior to placing it in the kiln, it will warp

less. This is especially important when you fire at high temperatures.

Removing a fired work from the kiln and placing it on a steel block for quick cooling is the type of action that causes uneven cooling, and can produce warping and cracking. When working on a very large formed piece, I sometimes allow it to *anneal*, or cool slowly, by placing it in another unheated or slightly heated kiln away from drafts.

If your piece warps during firing, you can flatten it with a steel plate or iron as soon as you remove it from the kiln. Quickly transfer the piece from the kiln to a very clean heat-proof surface, such as steel or transite board, and then immediately press it down with the plate or iron (photo A). If the

piece cools before you are ready to press it with the iron, don't do it or the piece will crack. Though you can purchase commercial steel plates, I had a blacksmith make one for me, and I use it for flattening larger work. To flatten small pieces you can use an old-fashioned flatiron with a nicely polished bottom surface. During the flattening process, the surface of the enamel may occasionally become marred. If this occurs, re-grind the piece or apply more enamel to your liking.

KILN FIRING STEPS

Most enamels applied to precious metals fuse between 1400° and 1500° F (760° and 815° C). If you use an indicating pyrometer, a device that measures the kiln temperature, you'll most often fire between those temperatures for two to five minutes. (As explained on page 49, however, if you're firing a transparent color directly on copper, you may need to use a hotter kiln.) Be cautious and aware that these firing times and temperatures are merely guidelines. (Pyrometers may be inaccurate. Kilns fire more quickly later in the day when they've been heated longer. Even the weather outside

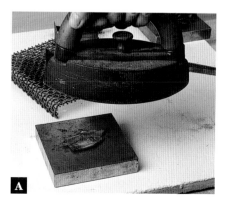

Linda Darty *Garden Candlesticks: Red*, 2003. 6 x 5½ x 5½ in. (15 x 14 x 14 cm) each. Copper, enamel, sterling silver. Photos by Henry Stindt.

can affect firing time!) With testing and experience, you'll learn how to fire intuitively with the enamels you use, the pieces you create, and the equipment you have.

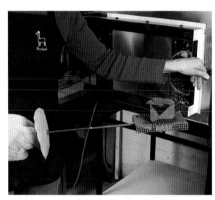

1. Position the completely dry, enameled piece on the trivet so only its edges are touching or so that the firing rack points are strategically placed. (How you place the piece on the trivet also affects how much the metal will warp when it expands and contracts. The more places a piece is supported, the less likely it will be to warp.) Place the trivet on a firing screen. It should be firmly balanced on the screen and not wobble (see photo). (If the trivet touches the enamel at any point, it will leave a mark during firing. Should this happen, you

can grind marks off with an alundum stone or diamond sanding sticks. Glass brush to clean, and then re-fire the enamel.)

2. Wearing a safety glove, use a firing fork to place the firing screen into the center part of the kiln chamber.

3. Leave the piece in the kiln for two to five minutes. After about a minute and a half has passed, peek through the door to see what the surface of the piece looks like. If you can't see the glass surface, you'll at least be able to see the color of the trivet or screen, and you can use that color as an indicator of whether or not the enamel has fused. Generally, a dull red to a cherry

red colored trivet indicates that the piece has reached the correct temperature for fusing. A hotter kiln glows more orange or yellow.

TORCH FIRING STEPS

Torch firing is another way to fuse enamel to metal. Acetylene, propane, or natural gas torches are all suitable for enameling. If fuel is combined with oxygen the torches burn hotter, and may be too hot for enameling. With practice, you can torch fire with just about every enameling technique. Mary Larom, who wrote a book on enameling in 1954, torch fired exclusively, even on controlled techniques, such as cloisonné and plique-à-jour. The procedure is quite simple. You use a flame to heat an enameled piece, usually from underneath, and then remove the flame once the enamel fuses. As I teach students, I encourage them to torch fire so they can watch the enamel closely as it progresses through the softening stages prior to fusion.

HOT TIPS
TRIVET & SCREEN MAINTENANCE

— Before using a trivet check its edges for any enamel residue that could stick to your piece.

— Periodically file the edges of trivets to keep them clean, and they will leave fewer marks on your work.

— When support screens sag or become less stable with age, gently reshape them with a wood or rawhide mallet.

Right: Kimberly Keyworth *Jubilation Bracelet*, 2003. 1 x 7 in. (2.5 x 17.8 cm). Sterling silver, 22-karat gold, enamel; fabricated, torch fired. Photo by George Post.

Far right: Deborah Lozier *Utility Brooch/City Street*, 1999. 3 x 2¾ x ⅜ in. (7.6 x 6.9 x .9 cm). Copper, enamel; torch fired. Photo by artist.

1. Create a trivet and stand to hold an enamel piece. This could be a firing screen on two firebricks as shown or a screen on a tripod. (Remember that heavy metals are heat sinks. If the trivets and stands you use are heavy, it will take you longer to heat the piece. If necessary, you can construct an aluminum can oven as shown in figure A or a brick oven around an enameled piece to hold in the heat. This is especially helpful when heating large pieces. Sometimes you may even want to use two torches simultaneously.)

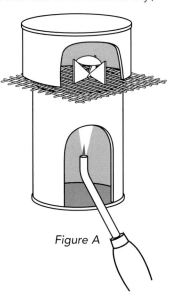

Figure A

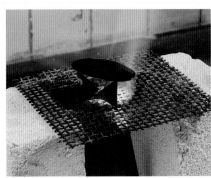

2. Place the enameled piece on the trivet and stand. Light the torch and use gentle overlapping strokes to move the flame underneath the piece.

SAFETY MATTERS

• *Good ventilation is especially important when torch firing because you'll be standing over the piece and may be breathing toxic fumes.*

• *Wear dark safety glasses with calobar lenses to protect your eyes when torch firing.*

FIRING STAGES

The same color used on the same form can create very different effects, depending on how thickly you apply it, and how hot and how long you fire it. You can learn to use over-firing and under-firing with control to create intentional rather than accidental surface treatments. First, however, you should learn to fire each enamel correctly.

A good way to get familiar with the way enamels flow and fuse is to fire pieces with a torch and watch the glass surface. You might also be able to see the surface if you move your head up and down, watching

HOT TIPS
TORCH FIRING
▬

— Torch firing can produce interesting results when the flame hits the enamel colors and causes them to oxidize unevenly. This special effect varies with different colors, and should be thoroughly tested.

— If you want to experiment with the color variations that occur when the carbon flame combines with the glass, move the flame to the top surface once the enamel reaches the orange-peel stage. Depending on the enameling technique you're using, this may or may not be advisable. Experiment! You can torch fire both lead-bearing and lead-free colors, but be advised that when the torch flame hits the surface of lead-bearing colors, they sometimes turn gray. Lead-free colors are not only more stable, but they are also safer to use when torch firing.

the reflections on the piece as you peek through the door of the kiln.

SUGAR-COAT TEXTURE

Within the first minute of firing a sifted enamel, the sugar-coat texture begins to form.

ORANGE-PEEL TEXTURE

As the glass softens and begins to fuse it forms a new texture, sometimes referred to as "orange peel," after the surface of the fruit skin.

FULLY FUSED SURFACE

A fused enamel surface is even and glossy.

OVER-FIRED SURFACE

On an over-fired piece, copper oxides may darken areas of the enamel.

CONTROLLED OVER-FIRING

With practice, you can create special effects on an enamel surface through deliberate and controlled over-firing. Fae Mellichamp's three necklaces pictured in the gallery on page 41 were all coated with the same clear enamel. The variation in color is the result of different firing times and temperatures.

Over-firing to Clear Transparent Enamels or to Create Copper Oxide Coloring

When over-firing most colors directly on copper, the color will change somewhat as the oxide is taken into solution with the glass. This might look like a green copper oxide halo or shadows of green, or perhaps dark black specks or black beneath some opaque enamels. In the case of transparent enamels sifted directly on copper, over-firing is used with control so that the copper oxide will be taken into solution with the glass, and the color will be more transparent (see Working with Enamel Colors, pages 48–50).

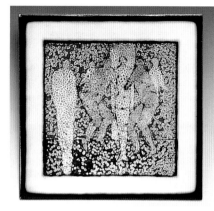

Break Up

When you over-fire one color on top of another, depending on the viscosity and softening point of the enamels used and the application amount, the top coat may break up, allowing you to see the bottom color bubbling through the top. The piece pictured above is from Harold B. Helwig's *Stone Stories* series. It is a good example of using controlled over-firing to intentionally create surface texture.

CONTROLLED UNDER-FIRING

If you carefully watch an enamel and remove it from the heat during the early firing stages, you can create some other interesting surface textures.

Sugar Firing

This under-firing technique is most successful if you completely fuse a coat of enamel underneath the desired sugar-fired layer. (Sugar-fired enamel will usually flake off if you apply it to bare metal.) Fire the base coat of enamel until it fuses and glosses. You can better control the sugar-fire surface by using uniform enamel particles, so sift out finer ones. Sift on the layer to be sugar-fired, and heat it only until you see the color of the powdered glass change while the texture remains gritty. This is a little

Right: Fae Mellichamp
Drum Bead Necklaces,
2004. 8 in. diameter (20.3
cm) each. Copper, enamel,
sterling silver. Photo by
Keith Wright.

Far right: Jan Baum
Seduction #1 and #2, 1998.
Various dimensions. 18-karat
yellow gold, 14-karat yellow
gold, stainless steel, enamel,
salt. Photo by Phil Harris.

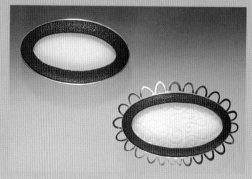

tricky and takes practice, but it creates a wonderful tactile surface as shown in Joe Wood's *Ero-Blue #1* on page 36.

Orange-Peel Texture

Firing an enamel only until it reaches orange-peel texture can give it an interesting finished surface as shown in Joe Wood's *Yellow Set #3* on page 36. Watch carefully, and stop heating the piece while it still has a bumpy surface.

SAFETY MATTERS

• *Cataracts can form after direct long-term exposures to high-intensity infrared rays. Calobar lenses protect your retinas from the infrared light rays emitted from a kiln or torch.*

FAST FACTS & HELPFUL HINTS

— Use a kiln or a torch to fire enamels. Both heating methods require ventilation.

— Base your firing time on the size of the piece, the enamel color used, the size of the support trivets and screens, and the location of the piece in relation to the heat source.

— Look at the color of the firing trivet. It can be a good indicator of when enamel is fused.

— Steer clear of using a timer to gauge firing. Successfully firing one piece for a certain time at a certain temperature does not mean this is the right combination for the next piece. Every enamel project and every firing is different—the color you use might have a lower fusing temperature; the size of the piece may be different; the trivet and screens may be heavier; the location in the kiln may be slightly different; and how long the kiln has been running might affect its temperature. Watch and check each piece until you're familiar with your kiln and learn to judge the temperature by the glow in the kiln or the color of the trivet.

HISTORICAL HIGHLIGHT

From its earliest beginnings, the firing process was basically the same as it is today. In the 12th century, Theophilus the Presbyter explained firing in his *Fiversarum Artium Schedula (On Divers Arts).* He wrote that the work to be enameled was placed in a flat, thin iron tray, and then covered with a shallow concave iron bowl perforated with holes. The cover had a ring on the top so it could be lifted. After placing the enameled piece in this iron container, Theophilus said one should:

> …heap up big, long pieces of charcoal and burn them strongly. Make a place among them and smooth it with a wooden mallet so that the iron tray may be lifted into it with tongs by its handle. Cover it and carefully set it in

position; then build up charcoal on all sides around and above it; and now take the bellows with both hands and blow from all sides until the coals blaze evenly. You should also have a whole wing of a goose or some other large bird, which should be fully spread and tied to a stick; fan the coals with this and blow vigorously on every side until you see between the coals that the holes in the cover are completely red-hot inside; then cease blowing. After waiting about half an hour, uncover it gradually until you have removed all the coals. Again wait until the holes of the cover grow black inside; then lift the tray out by its handle; and, while keeping it covered, put it in a corner behind the furnace until it is completely cold.

Cleaning Metal & Enameled Surfaces Between Firings

The method you use to clean uncoated metal or enameled surfaces between firings will vary depending on the metal you're using, the affects you want to achieve, and whether or not you've touched an enameled surface. For cleaning uncoated copper edges use a metal file, sandpaper, or an alundum stone. To clean an enameled surface, use a clean glass brush under running water.

COPPER FIRESCALE

A

B

After firing enamel on copper, any areas that weren't covered with enamel will be covered with *firescale*, a brownish-red oxide that forms on metal when it's exposed to high heat (photo A). Firescale flakes off as the metal cools (photo B). You can throw it away or collect it to use for special enameling effects. (See page 102 for further information.) If you plan to apply an opaque color over copper with

Amy O'Connell *Necklace for Sun Worship*, 2003. 14 to 16 in. (35.6 to 40.6 cm). Copper, sterling silver, enamel, garnet, amber, carnelian, yellow opal. Photo by artist.

firescale, you can simply brush it off and rinse the metal. (Be certain the water sheets across its surface, indicating it's clean.) If you plan to apply a transparent color over an uncoated copper surface, you may want to selectively or completely remove the firescale, or use it in your design.

In these samples, clear enamel was fired over a firescaled surface (left) and over clean copper (right).

PICKLE BATH

C

If the enamel you're using on the first side of the copper is acid-resistant, you can place it in a pickle bath of sodium bi-sulphate or in a dilute solution of nitric acid (3 to 6 parts water to 1 part acid)

for a few minutes until the exposed copper is restored to its original color (see photo C). Most lead-free colors have high acid resistance, so even if I'm planning to use lead-bearing colors on the front side of a piece, I'll often coat the back side with lead-free color so I can use pickle or acid. Some lead-bearing colors are also fairly acid-resistant, but you should always test before pickling. If the colors are not acid-resistant, the enamel surface will be altered after pickling or during subsequent firings, taking on an uneven matte finish. After pickling, rinse metal well.

BALL CLAY

D

If you don't wish to clean metal with pickle, you can coat Scalex or ball clay on the side of the piece that won't be covered with enamel prior to firing (see photo D). The mixture, which is available commercially, paints on easily with a brush, is a firescale inhibitor and simply peels off the metal once cool (see photo E). After painting ball clay on

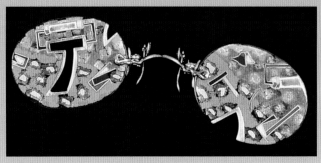

Right: Gretchen Goss
Family Reunion: Madlene, Cuba, 1997. 12 x 10 in. (30.5 x 25.4 cm). Copper, enamel. Photo by artist.

Far right: James Malenda
Tony's PC, 1999. 2½ x 5 in. (6.4 x 12.7 cm). Enamel; champlevé, cloisonné. Photo by Jeff Unger.

E

metal and letting it completely dry, make certain it has not contaminated the metal surface you'll be enameling. If so, wipe the surface clean before applying.

DUAL-SIDED FIRING
You can apply an enamel coat to both sides of a metal piece at one time, and fire them simultaneously to prevent firescale. This method also reduces the risk of the metal warping. This treatment is easiest to accomplish with liquid enamel. (Refer to page 104.) Alternately, you can paint or spray gum binder under and even over the one sifted side before turning the piece over to sift enamel on the opposite side.

CLEANING THE EDGES
Even with both sides of a copper piece coated with enamel, the metal edges will be covered with firescale after firing. Firescale can "jump" off the edges during subsequent firings, creating black specks on the front of a piece. To prevent this, simply clean the

F

metal edges between firings with a file (see photo F), emery paper, or underwater with an alundum stone. Always file or sand away from your enameling area so you won't contaminate the workspace.

FINE SILVER, GOLD & STERLING SILVER
Firescale does not appear on the surface of fine silver or fine gold during firing. When working with sterling silver and gold alloys containing copper, you should prepare the metal so no firescale forms on these surfaces. (For further information on depletion gilding, see pages 23–27.) After firing enamel on one side of any of these metals, if you only handle the piece by its edges, you can simply rinse it under running water before applying the enamel to the other side. If you do touch the surface of the glass, clean it as described in the next section.

CLEANING BETWEEN COATS OF ENAMEL
If you touch a fired enamel surface, such as when filing metal edges, simply clean the glass under running water with a fiberglass brush (also called a *glass brush*) until the water sheets evenly across its surface. Do not use ammonia or other detergents with the glass brush. If you use soap on the brush and later clean a cracked piece, the ammonia or soap could get into the crack and cause that area to remain cloudy after firing.

REPAIRING CRACKS
If you discover a crack in a fired piece, either as a result of dropping and breaking or as a firing defect, resist the temptation to touch it before re-firing. It's difficult to heal a crack if oil from your skin is deposited in the crevice. You may want to rub it with your fingers as you think, "oh my, there's a crack." Don't do that! If you immediately put the piece back into the kiln, you'll have a better chance of thoroughly healing the crack.

If you do touch a cracked enamel (or your friend does), what do you do next? Simply use a clean glass brush under running water to wash the piece well, and then let it completely dry. As mentioned above, do not soak the piece in ammonia and water or use any kind of detergent.

Above: Jessica Turrell *Axe Head Brooches,* 1995. 2 x 1½ in. (5 x 3.8 cm). Sterling silver, gold, enamel; cloisonné. Photo by artist.

Left: Sarah Perkins *Cherubfish Teapot,* 2003. 4 x 4½ x 3 in. (10.2 x 11.4 x 7.6 cm). Silver, amethyst, enamel. Photo by Tom Davis.

Removing & Grinding Enamel

Let's assume you don't like the color you've applied to an enamel, or perhaps you don't like it in one particular area. How often have I thought, "I liked that so much better before I added that last color!" What can you do? How can you remove the enamel? The same question may arise if you've let a piece dry on a table that's covered with firescale—suddenly, after firing, you see black spots on the enamel. No matter what the reason, don't worry. There are ways you can remove enamel.

Some of these methods are the same ones you'll use to level an enameled surface when employing champlevé or cloisonné techniques. When using these techniques you'll inlay color in layers, building up the enamel either to the level of cloisonné wires or to the raised surfaces of champleve. In order for these surfaces to be smooth and even, you'll need to level the surface of the fired glass.

METHODS FOR REMOVING & GRINDING ENAMEL

There are tools and techniques for removing enamel in both small and large areas. After grinding, always glass brush before refiring.

Alundum Stone

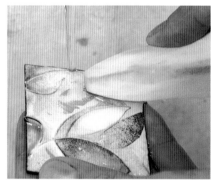

This stone is made of aluminum oxide, and it cuts and levels glass very quickly. You can purchase alundum stones from some commercial enamel manufacturers. The stones are available in various grit sizes, such as 150, 220, and 320, with the higher number indicating the finest stone. The lower number, or coarser stone, cuts more quickly but leaves more scratches on the enameled surface. Always use alundum stones under running water so the aluminum oxide residue is rinsed off the piece as it is stoned and not ground into the glass.

Diamond Sanding Sticks & Paper

You can purchase commercial diamond sanding sticks from enamel suppliers and metal tool companies, or you can buy diamond sanding paper and glue it onto sticks or sponges. I like these abrasive tools because they cut quickly and leave fewer marks on the enameled surface than alundum stones. They come in various grit sizes with the higher number indicating the finer grit surface. See photo C, page 52. If you don't wish to re-fire an enamel to a glossy surface, you can also use diamond sanding sticks or paper in order from the coarsest to the finest and produce a nice matte finish.

Diamond Burrs

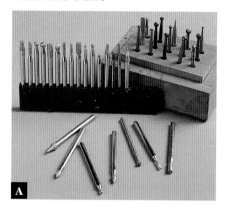

Different shapes of diamond-impregnated burrs (photo A) are incredibly helpful when you want to remove color from a small inlaid area or on an area that's within metal walls, such as on a cloisonné or champlevé piece (photo B). You'll use these burrs in a flexible shaft machine, and to ensure that they last longer, always place a drop or two of water on the surface of the glass. Sharp pointed diamond burrs are great for removing a speck of unwanted firescale and for drilling holes through fused enamel.

Burrs without diamond-impregnated surfaces, like the ones in Photo A on the right, are not used to remove enamel. I use them to texture or remove metal.

Sandblaster

A sandblaster quickly removes enamel, and you can use one to create a concrete-like or matte surface. If you have access to a sandblaster, experiment with different grits of sand or bead blasting. Always work with appropriately ventilated equipment and wear a safety mask. You should thoroughly clean a sandblasted enamel with a glass brush before it is re-fired.

HOT TIP

—No matter what method you choose to level or remove enamel, always remember to completely clean the surface with a glass brush under running water before applying subsequent layers of color or re-firing the piece.

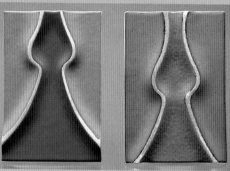

Top: Susan Remnant *Big Rings*, 2003. 1⅗ x 1³⁄₁₀ x 1⅕ in. (4.2 x 3.4 x 3.2 cm). Sterling silver, fine silver, copper, enamel. Photo by artist.

Center: Becky Brannon *Spire Brooches*, 2003. 2 x 1 in. (5 x 2.5 cm). Fine silver, enamel. Photo by artist.

Bottom: James Doran *Table Stakes*, 1996. 4 x 10 in. (10.2 x 25.4 cm). Copper, copper foil, metallic lustres, glass marble, stainless steel wire; electro-formed, carved, under fired, etched. Photo by artist.

Left: Helen Elliott *Marked Moments 2,* 2002. 6 x 6 x ¾ in. (15.2 x 15.2 x 1.9 cm). Steel, wood, porcelain enamel; limoges. Photo by Greg Staley.

Far left: April Higashi *Ruby & Emerald Cuff,* 2001. 3 x 7½ x ¼ in. (7.6 x 19 x .3 cm). Silver, tourmaline, enamel. Photo by George Post.

Right: Linda Darty *Fish Server,* 2003. 12½ x 4½ in. (31.7 x 11.4 cm). Sterling silver, fine silver, enamel. Photo by Robert Diamante.

Far right: Barbara Seidenath *Crystalline,* 2002. 1½ x 1¾ x ¾ in. (3.8 x 4.4 x 1.9 cm). Sterling silver, enamel. Photo by TBD.

Making Color Tests

Enamel colors look very different in the jar than they do fired. In order for you to gain familiarity with and control in using individual enamels, the importance of making color tests cannot be overemphasized. As an example, you create many colors from one transparent hue. The different enamel colors depend upon the base metal used, how thickly the enamel is applied, and how hot it's fired. Opaque colors will vary too, according to how thickly they're applied and how long or hot they're fired.

In my studio I test enamel colors in many different ways, and I store them for future reference, preferring to make color tests on small metal pieces. To prepare the metal, a hole is drilled at the top of each piece so it can hang on a hook on a large sample board. When working out a design, I can easily remove metal samples from their hooks and place them together in different color combinations. Even though I may have used a color for 20 years, I still frequently refer to its test sample. The following are ways of making color tests that work for me, but keep in mind that you can plan and test your enamels any way you like.

TESTING TRANSPARENT COLORS ON COPPER, WHITE ENAMEL & FOILS

1. Cut several copper rectangles that are each at least ¾ x 2 inches (1.9 x 5 cm).

2. Drill a small hole in the top of each rectangle.

3. Counter enamel the back side of the copper.

4. Test transparent enamel colors directly on the copper, over clear enamel, over opaque white enamel, over gold foils, and silver foils.

TRANSPARENT COLOR TEST

1st Firing	2nd Firing	3rd Firing	4th Firing
○ Transparent directly on copper	○	○	○ Transparent color
	White		over
	Clear enamel	Silver foil	the
	Clear enamel	Gold foil	entire
	Clear enamel		piece

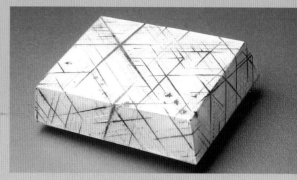

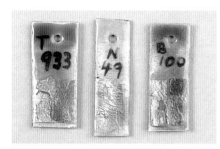

TESTING TRANSPARENT COLORS ON STERLING SILVER

Enamel colors look slightly different on sterling silver than they do on fine silver, so I recommend making separate color tests. I often purchase colors just for working on silver, and this is how I test them.

1. Cut very small pieces of 18- or 20-gauge sterling silver, each approximately ½ x 1½ inches (1.3 x 3.8 cm).

2. Drill a small hole in the top of each piece.

3. Following the method described on pages 26–27, heat and pickle the sterling silver pieces multiple times, raising the fine silver to the surface.

4. First test the colors on the front side of the metal before counter enameling. This way, you won't risk firescale coming to the bare uncoated front surface before you cover it with glass. Test the transparent colors directly on the ster-

ling silver and over clear enamel. Also test the enamels over gold foils and silver foils if needed.

TESTING OPAQUE COLORS

You can test opaque enamels simply by sifting them on drilled copper. You'll want to test opaque colors at different firing temperatures, so make the samples on rectangles that are each approximately ¾ x 2 inches (1.9 x 5 cm). One way is to test each opaque color fired to maturity, over-fired, and sugar-fired.

TESTING COLORS FOR SPECIFIC PROJECTS

Even if you create many sample boards, you may still need to test colors for specific projects to see

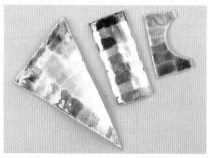

how certain enamels look in relation to each other or how they blend together. I usually use a scrap piece of silver or copper, drill it, and then paint on the colors in strips or blends. The samples above were made for my fish server pictured in the photo, top left.

TESTING OTHER MEDIA

Pre-enameled steel tiles are excellent to use when you need to make several color tests together on a white background. I test painting enamels, ceramic pigments, enamel marking crayons, watercolor enamels, and acrylic enamels on pre-enameled steel tiles.

OPAQUE COLOR TEST

1st Firing	2nd Firing	3rd Firing
○	○	○
Sift opaque color		
over entire piece	Sift bottom two sections and	
and over-fire	fire normally	Sift bottom square only and under-fire

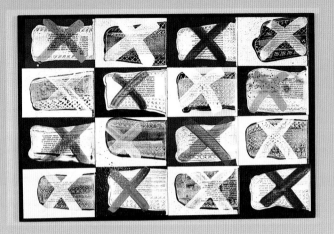

Working with Enamel Colors

Whether it's picking out what to wear, choosing what appliance to buy, or even ordering checks—everyone makes color decisions on a daily basis. Color has a basic, instinctive, visual appeal, and though the study of color theory can seem complex, there are really just a few basic things to remember that will help you feel confident when you make your enamel color selections.

If you feel insecure about working with color, look at your surroundings: the clothes you wear, the way you choose rugs or home furnishings, curtains or wall paint. Study the subtleties of the browns and greens in leaves, looking hard enough to see the yellows, the pinks and the reds as well. See how

Figure A

the rocks relate to each other in a river, or the way the flowers burst into a blaze of color in a carpet of mossy green grass. Pick up a book on quilts and look at how textile artists choose the fabrics they place together and ask yourself why you like certain ones best.

Color is a product of light. As the light changes, the color changes. Think about the color of grass on a summer morning as the sun rises, at noon when it's so very bright, and then in the evening as the sun fades. The grass is definitely green. But, if you were trying to paint that color with enamel, you'd either be choosing different greens, or layering other transparent colors over the green to darken or lighten it.

Colors also change according to their surroundings. Even in the same light a color will appear different depending on the colors that are adjacent to it. As shown in figure A, the same pink square looks different when placed on larger squares of other colors. Cool colors (blues, greens, and purples) will recede in a composition while warm colors (reds, oranges, and yellows) advance, giving the piece depth and volume.

As you work with color, it's helpful to hold a piece away from you and

think about how your use of color makes your eyes move around a composition; how color creates balance, harmony, or emphasis in one area. Your piece may need a focal point that creates interest and gives it life. If the piece is primarily lavenders and purples, you might want a touch of gold somewhere to give it a little sparkle. (The complement of purple is yellow.) A little orange spot in a piece that's predominately blue or some red in a field of green might help your composition. Give your color choices ample consideration and use them judiciously. It's tempting to want to use every single color you have in your palette in every piece you make! Beautiful colors of glass are seductive. It's easy to end up with a piece that looks as if you've gone into a candy store and spilled all the jellybeans! By limiting your palette, you'll establish an overall mood in your composition. Using related values and subtle color changes might be all your piece needs.

Different enamels with similar expansion and contraction properties can be mixed in a paint tray or jar and used as one new color. If you're using opaque colors this might result in speckled-looking enamels because you'll likely see the grains of glass in the mix.

You'd certainly need to test new mixtures to know what colors they create. I achieve more control if I use a paintbrush to mix colors directly on the piece.

Another way of working with color, is to mix it visually rather than physically. In the 19th century, pointillist painters used this technique by applying small dots or points of color next to each other to alter color sensations. The trees in a painting may have looked green, but actually were made up of many different hues to make that green come to life. There might be yellows, browns, blues, even lavenders and pinks in the color that we instinctively see as green. The blue/green enamel you see on the fish server on page 47 is full of gray, pink, yellow, and lavender.

When planning an enamel and deciding which colors to use, it's helpful to use colored pencils, markers, watercolors, or any other paint on paper to plan your design. I'll use tracing paper and draw my piece several times. This way I can try different color combinations before I begin making my enamel tests. Some projects are spontaneous, and this color planning may not be necessary, but in others, especially those involving complex metalwork, advance preparations pay off.

After designing on paper, you can make more specific color tests on metal. If you've already made color samples, simply hold them together and see how the colors look. If you're going to be wet inlaying, shading, or sifting colors that blend or overlap, you may want to make quick tests on scrap metal.

HOW TRANSPARENT ENAMEL COLORS REACT ON DIFFERENT BASE METALS

In addition to understanding color, it's also important to know how enamel colors react to different types of metals. Opaque enamel colors are very simple to use. If fired correctly, they should look almost the same on any base metal. Transparent colors react differently to each base metal. Seeing through them to the base metal itself adds yet another layer of visual information.

Transparent Enamel Colors on Copper

Some transparent colors on copper will be clear and brilliant if fired correctly. Others, because of their composition, will need a layer of properly fired clear flux beneath them or they will look dark and muddy. You should experiment and make samples to test transparent colors on copper.

These samples show a transparent orange enamel that does not fire to clarity directly on copper, next to the same transparent orange when it is fired over clear flux. This is a color that would need a clear undercoat.

These samples show clear flux that is underfired (left), and fired hotter, to clarity (right). Transparent colors applied over clear flux will be more

brilliant if the flux is first fired so that it is clear.

Firing Transparents to Clarity on Copper

When firing transparents to clarity on copper, the firing should be hot or long enough for the enamel to absorb the copper oxide and take it into solution with the glass.

Method One

Fire the transparent colors directly on copper at 1400° to 1500° F (760° to 815° C) for 3–4 minutes, two or more times, until the color is clear (as in the samples above showing multiple firings of ivory beige directly on copper).

Method Two

Fire the piece in a hotter kiln, 1600° or 1700° F (871° or 926° C), in only one firing. (This method is not a good choice if you have solder joints that might detach or become brittle at high temperatures.)

Method Three

Fire the transparent enamel with a torch, holding it underneath the copper past the point that the enamel has fused, until you see it take on a bright orange color from heat. Repeat the torch firing as needed until the enamel is clear.

Transparent Colors on Fine Silver or Sterling Silver

Transparent enamel (clear flux) for silver is different from that used for copper. These two distinct materials should be purchased and stored separately. You'll only fire transparent

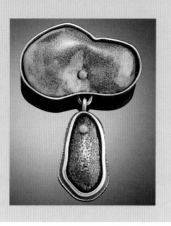

Left: Deborah Lozier *Useless Beauty/Funnel*, 2002. 7⅝ x 5¾ x 5¾ in. (19.4 x 14.6 x 14.6 cm). Copper, enamel, patina. Photo by artist.

Far left: Felicia Szorad *Blue Dangle Brooch*, 2002. 3 x 2 x ¼ in. (7.6 x 5 x .6 cm). Sterling silver, copper, enamel; torch fired. Photo by Taylor Dabney.

Right: Falcher Fulsager *Creation 5*, 2003. 4 x 4 in. (10.2 x 10.2 cm). 18-karat gold, enamel: 24-karat gold, garnets, diamonds; cloisonné. Photo by Jael Fulsager.

Far right: Marilyn Druin *Red Circle*, 2000. 1⅝ in. diameter (4.1 cm). Fine silver, 24-karat gold, 18-karat gold, enamel, diamonds; cloisonné, guilloché, basse taille. Photo by Bob Barrett.

enamels on silver to the correct softening point. Never fire transparent enamel on sterling silver or fine silver as hot as on copper. Over-firing to absorb copper oxides doesn't work the same way with silver alloys. (Firescale on sterling silver must be depletion gilded before enameling begins. Once sterling silver is properly prepared, and if not over-fired, it will react with enamel in most cases like fine silver does.)

HOT TIP

— Firing transparent colors directly on copper takes a lot of heat, but don't get into the habit of over-firing all enamel. Fire successive layers of transparent enamels at normal fusing temperatures and times—just until the glass is fused. It's only on the first firing that copper oxides at the interface of the base metal need to be absorbed into the glass.

Transparent Colors on Gold

Just as you can see the cool white of silver beneath transparent enamel colors, you also can see the warm yellow of gold beneath transparent colors on gold. When in doubt about how a transparent color will look on different metals, run a test! (Refer to pages 46–47 for further information.)

The same transparent blue enamel fired over silver foil (left) and over gold foil (right)

Three colors that look brilliant when fired over gold foil

LAYERING COLORS

When considering how transparent enamel colors will look when fired, it may help to imagine how flat sheets of stained glass look in a window or held up to a light. When you hold a yellow sheet in front of a blue sheet, depending on how dark the colors are, the two panel colors can combine to look green. As you look through the yellow, you still see the blue. If you look through a pale pink glass sheet to dark blue glass, the dark color stays dominant, and depending on how dark it is, the blue may or may not be affected by the pale pink. If, however, the back sheet is pale blue and the front sheet is pale pink, the pink sheet has more of an effect on

the combined color, and you will see lavender. It takes practice to work with transparent colors with ease. This is yet another reason to limit your initial palette of enamels. With practice, you'll become more familiar with each color by experimenting with it in a variety of combinations. It's not necessary for you to repeat the same color in every fired layer, and it's often more interesting if you use different hues in subsequent firings. Bill Helwig, my first enameling teacher, once told me, while I was creating a seashell image, to fire transparent lime green over a coat of pink. I resisted, wondering how in the world lime green could make my pink seashell look better. He said that red and green are complementary colors. I knew this principle well, but I had never associated it with layering transparent enamels. It was a startling discovery, and one I've never forgotten. I was delighted to see that pink seashell still look pink, but suddenly be more luminous, rich, and almost iridescent with transparent lime green added to its surface. I soon began sifting oranges over blues, yellows over purples, and layering many variations of these combinations. I encourage you to experiment. Depending on the darkness or lightness of the enamels you use, the effects can be dramatic.

SHADING & VALUE GRADATIONS WITH SIFTING & PAINTING

In addition to mixing colors, creating new ones, and achieving interesting visual effects through layering, you also can combine and shade colors in a single layer. Consider using sifting techniques and slightly overlapping one color with another. If you thinly apply the enamels, once fired the lower color will show through the top color. If you want to paint with enamels, you can apply two or more wet colors and use a brush to mix them together where they meet. This method of color blending is not so much a painting technique as it is a mixing technique. You'll use your paintbrush in an up-and-down motion to push the grains of glass around each other into position. When they dry to a powdery finish prior to firing, your two blended enamel colors may not look as good as they did wet. Once fired, however, they should be fine. (For further information, see Painting with Enamel on page 88.)

be fine. (For further information, see Painting with Enamel on page 88.)

HOT TIPS
TRANSPARENT COLORS ON SILVER

—Always fire clear transparent enamel under warm enamel colors, such as yellows, reds, oranges, pinks, and even some purples and lavenders. Because glass is slightly alkaline, it creates silver salts when applied over silver and silver alloys. These salts are taken into solution with the glass, discoloring warm colors. If you want to achieve optimum clarity and brilliance when working with warm colors, fire gold foil beneath them. (For further information on firing foils, see page 80.)

—Though there are some transparent red enamels that look nice over silver flux, most are more beautiful with a gold foil layer under them.

—Cool transparent colors (blues, greens, grays, turquoises, teals, and some purples) are beautiful when fired directly on silver without a base coat of fired clear enamel.

—Be very careful not to over-fire sterling silver. Even after the fine silver has been brought to the surface, brown spots may appear through the transparent colors as copper oxides are taken into solution with the glass.

FAST FACTS & HELPFUL HINTS

— **Hue** simply refers to the name of the color. Red is a hue, yellow is a hue, and orange is a hue.

— **Value** refers to the lightness or the darkness of a color.

— **Intensity** refers to the brightness of a color. Because a color is at full intensity only when pure and unmixed, there is a relationship between intensity and value.

— **Warm Colors** are those colors related to the sensation of warmth, such as yellows, reds, and oranges.

— **Cool Colors** are those colors related to the sensation of cold, such as blues, greens, purples, and lavenders.

— **Primary Colors** are pure and exist without being mixed. There are only three: red, yellow, and blue.

— **Secondary Colors** are mixed using only primary colors. The secondary colors are purple (a mix of red and blue), green (a mix of blue and yellow), and orange (a mix of red and yellow). These are the only secondary colors.

— **Complementary Colors** are opposite from each other on the color wheel.

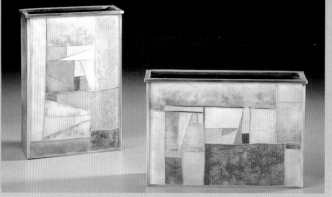

Finishing Enamels

An enameled piece fired to fusion has a glossy glass surface. This may or may not be the surface quality you desire on a finished piece. In the section Firing Enamel, you learned how to control the appearance of the enamel surface through firing times and temperatures. In this section, you'll learn how to use abrading techniques or chemicals to create interesting enamel surfaces for a finished artwork.

ETCHING PASTES & SOLUTIONS

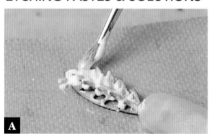

A

Chemical etching pastes and liquids are available from enamel suppliers, stained glass suppliers, and many craft stores. To use the thick etching paste, simply paint it on the glass surface (photo A), leave it there for the amount of time the manufacturer suggests (usually 5 to 10 minutes), and rinse it off. (Make sure to follow the directions on the bottle.) Because you dip the entire piece into liquid, an etching solution is more convenient for some shapes. Depending on the thickness of the enamel and the strength of the liquid, an etching solution may be

able to remove all of the enamel from a metal surface if left to soak overnight or longer.

ABRASION TECHNIQUES

B

Occasionally, you may want to create a level surface by grinding an enamel until it's flush with a raised metal surface. This is often the case when you're employing champlevé or cloisonné techniques. The most efficient way to accomplish this is by using an alundum stone (see photo B) or a diamond sanding stick (photo C). (Refer to page 44 for further information on this method of removing enamel.) Whether the surface of an enamel is to be level or not, there are several ways to create a matte finish using similar techniques.

HAND-POLISHING TECHNIQUES

To level the surface of an enamel, gently grind it using an alundum stone or a diamond sanding stick under running water (photos B and C). If you don't plan to re-fire the piece and want to hand-polish it to a matte finish, use the finest stick or stone that will adequately do the job. I find diamond sandpaper works most efficiently and creates the fewest marks on the glass surface. (For greatest efficiency when hand-polishing, I start with nothing more abrasive than 320-grit diamond paper or sticks. The heavier grits leave more marks that will later need to be removed from the piece, and they can cause pitting.)

C

Once the surface is smooth, continue working with the next finest grit until all the marks from the previous grit disappear. Before changing grits, however, always clean the enamel surface with a glass brush to remove all sanding residue (photo D). If you don't clean between grits, you might continue

scratching the piece! Continue working under water as you progress to the finest grit and create the surface you desire.

Once you've ground an enamel with the finest grit diamond paper, you might like the surface as it is, or you may want to further refine it with wet emery papers (photo E). Begin with the finest grit emery paper that will effect the glass, and then work up to a 600- or 800-grit paper for a final polished finish. Micron papers with a very fine 1000 grit are also available, and you can use them if you wish. (Remember to always glass brush the enamel before switching grits!)

If you want even more luster, you can rub an enamel piece with a paste of cerium oxide and water

and/or finish it with microcrystalline wax or paste wax (photo F).

It's also important to finish the edges of an enamel if they will show in the final piece. Sand the edges with various papers, working from 320- or 400-grit to 600-grit, and then polish them with steel wool or a buffing wheel.

MACHINE-FINISHING

A wet rubber drum sander is a machine with a motor, a shaft, and at least one rubber drum attached to the shaft that can be fitted with sanding belts. The sander is used wet, usually with water dripping down onto its belt when running. The water rinses sanding residue away while a piece is being polished. The

rubber acts as a cushion, making it easy to control the pressure of a piece against the sanding belt. Wet rubber drum sanding belts come in various grit sizes with the higher number indicating the finer grit. To sand an enamel using this machine, carefully hold it against the lower third of the wheel (photo G). Use the finest grit sandpaper that will do the job first, working from 320- or 400-grit paper up to 600- or 1000-grit paper for the

HOT TIPS
HAND-POLISHING SHORTCUTS

— Method One: Level the enamel surface with a 320-grit alundum stone or a diamond stick. Clean the piece well with a glass brush. Re-fire the enamel to "heat-polish" it. Begin sanding again, this time with a fine 600-grit wet emery paper. (After firing, you may find that you need to sand with the 400-grit paper first, glass-brush the piece, and then move to the 600-grit paper.) Finish as desired, following the methods described in this section. Note: this timesaving technique only works if the stoned piece is re-fired to a completely level surface.

— Method Two: Level the enamel surface, clean the piece, and re-fire as described above. Then, following the manufacturer's instructions, paint the enamel

surface with a heavy coat of glass etching cream or dip it in a glass etching solution. Once the surface becomes matte, thoroughly clean it with a glass brush. If you wish to create more luster, sand the enamel surface with 600-grit wet emery paper.

—Note: Some sifting and limoge techniques do not require a level enamel surface. This means you can skip the sanding step and simply paint the surfaces you want to be matte with the glass etching cream or dip the piece in a glass etching solution. If you desire more luster, you can sand with wet 600-grit emery paper after etching, but be aware that this will only polish the high spots on the enamel, and the finished surface may not look even.

final finish. (Because I like to work slowly, sometimes I start the entire process with 600-grit belts!) Take great care not to abrade too much color too quickly. Because sanding belts cut so fast, it's easy to lose the last few firings of enamel color.

Here is one way to machine-finish:
1. If you're new to the equipment, you may want to fire at least one layer of well washed clear enamel over the final colors so you won't lose the colors you like.

2. Use a well-worn 320-grit belt if you want to quickly abrade the glass down to the wires or metal. If you prefer to work more slowly, begin with a 400-grit belt or even a 600-grit belt.

3. Use an old 600-grit belt to smooth the glass and remove all scratches from the previous belt. You also can round the edges of the glass if you like.

4. Use a 1000-grit micron belt to achieve a higher polish and to remove all the scratches from the 600-grit belt.

5. Depending on the degree of sheen you want for the final finish, you may want to buff the piece with crocus compound, cerium oxide paste, or microcrystalline wax. Wipe the enamel well with a soft cloth to remove any residue of wax or buffing compound.

HOT TIPS

— If a pit appears in a polished enamel, look at it carefully to determine if you should drill out the pit with a diamond burr or simply fill it with a grain or two of glass. If the defect is a bubble, you'll need to drill or abrade it to allow the air to escape. If the pit is a dark firescale spot, use a diamond burr to remove it, glass brush, and refill the pit with glass.

— If you finish a surface with a very fine grit sanding stick or paper before re-firing to gloss, the enamel will be less likely to have pits in its final surface. If there are deep scratches in the enamel when it's re-fired, it's more likely that the enamel surface will be uneven as it reflows in the final firing.

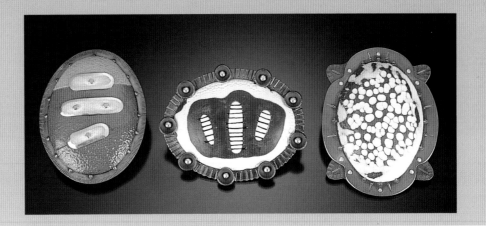

Right: Jesse Bert *Brooch Series.* 3 x 2 ½ x ¼ in. (7.6 x 6.4 x .6 cm). Silver, copper, bronze, enamel. Photo by Robert Diamante.

Left: Doug Harling *Green Garden,* 2002. 2¾ x 2 x ¾ in. (7 x 5 x 1.9 cm). 22-karat gold, silver, enamel; granulated, fabricated, cloisonné. Photo by artist.

Far left: Joe Wood *Torch Series Brooches,* 1995. 2½ X 1 X ½ in. (6.4 x 2.5 x 1.3 cm) each. Enamel, copper, sterling silver. Photo by Dean Powell.

Setting Enamels

Since, in most cases, you may not want to reheat a finished enamel, setting it as a piece of jewelry can be challenging. Though soldering a finished enamel is possible (see pages 22–23), soldering before enameling or using cold connections after enameling are two setting solutions that are less likely to damage the glass. Before even beginning to apply glass to metal, carefully plan the mounting or setting. For example, after the enamel is completed, it would be frustrating to realize that you should have drilled holes in the piece for decorative rivets or screws or that you should have designed a simpler shape for a bezel setting. Although I may sometimes use a bit of glue (most often silicone rubber sealant) to hold an enamel in place, I don't suggest this as your primary means of setting an enamel. I always prefer that an enameled piece be held in place with a mechanical device such as prongs, screws, rivets, or bezels. Enamels will far outlast the glue that bonds them to metal. (Remember, there are enamels that date back to the 13th century B.C.!)

HOT TIPS
SOLDERING WITH ENAMELS

—When soldering on the back of a finished piece there should be no counter enamel on the piece where the solder will flow. (You can grind it off with diamond burr, or not apply it at all, in the place you wish to solder.) With the saw and solder technique for champlevé enameling, counter enamel is not necessary on small pieces so they are simple to solder as a final step (see pages 120 and 163–166). Using hard solder is usually best so that you re-flow the enamel completely, essentially torch firing it. (Medium or easy solder may flow at a temperature that is just hot enough to cause cracks or pits in the enamel, though I have used easy solder with a tiny torch tip, for a very fast solder operation.)

—Use medium or easy solder to "kiln solder" or to solder the piece in the kiln, during a firing. The pieces to be soldered should be set up on a trivet or piece of mica, and arranged so that there is a clean, tight fit in the place where the joint will be. Apply soldering flux to the area where the solder should flow (see photo above), being careful that it will not contaminate the enameled surface. Fire the piece in the kiln as if fusing glass. In fact, I usually do this during a normal

firing while I am fusing enamel in some other area of the piece.

—Use eutectic, IT, or hard solder if the soldering operation will be done before the piece goes in the kiln or is torch fired. I find that hard solder holds up very well for the vessel forms I make and I like it better than IT because it is easier to work with. I also find it to be less brittle than eutectic solder and less likely to come apart if the piece gets too hot. Keep in mind that the enamel will pit or discolor if it is applied over hard or IT solder because they both contain zinc. I always use eutectic solder that is made of only copper and silver, if I plan to enamel over the solder joint. If using transparent colors over eutectic solder on silver, it should be depletion gilded until the surface is bright silver before enameling.

Above: Kristin Mitsu Shiga *Nebula Brooch*, 1999. $2^1/8$ x $2^1/2$ x $3/4$ in. (5.3 x 6.4 x 1.9 cm). Fine silver, sterling silver, enamel, hair; fabricated. Photo by Courtney Frisse.

Left: Sarah Krisher *Lost and Found in the 35 Acres*, 2001-2002. $6^1/2$ x $4^1/2$ in. (16.5 x 11.4 cm). Copper, enamel. Photo by Robert Muller.

COLD CONNECTIONS

An excellent way to set a finished enamel or add enameled components to an artwork is with cold connections. It's always a challenge to create a setting that's unique and fits a particular piece, but once you understand the techniques explained in the following examples, many variations are possible.

Sarah Turner used tube rivets to set the belt buckle in photo A. Photo B shows the back side of the rivets on Jesse Bert's brooches that are pictured on page 55.

Setting Small Enamels as Jewelry or Objects

If you're enameling on a small scale and wish to make cold-connected metal settings, you'll need some basic metalworking skills, such as using a jeweler's saw, and drilling with a hand drill, drill press, or flexible shaft machine. If you don't already know these techniques, I suggest you find a good book that can teach you these beginning metalworking skills.

Tube-Riveting Enamel

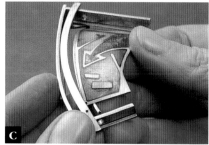

A functional rivet is a small metal rod, in photos C and D a hollow tube, slid through two drilled holes, one in each of the pieces to be joined, and then flared where it protrudes. The flared ends prevent the tube from sliding back through the holes. I like using tubing rivets so I don't have to hammer near the surface of a finished enamel.

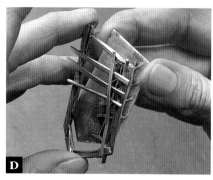

You can solder the tube to the back of the metal before it's enameled and set it from the back side, or the tube can protrude from the front, and you can flare it on both sides. It's relatively simple to gently flare a piece of tubing using a pointed scribe or a smoothly sanded nail head. You also can solder a decorative head onto one side of the tube (photo E). If you choose this option, the tube will only need flaring on one side.

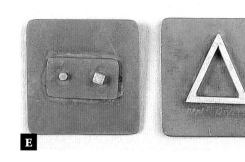

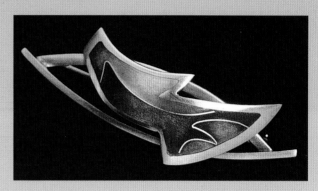

Right: Sun Hwe Park
Shadow C Brooch, 2000. 2½ x 2 in. (6.4 x 5 cm). Sterling silver, enamel, tourmalines. Photo by Mark Johnston.

Far right: Linda Darty
Garden Brooch #2, 2000. 1 x 1¾ x ¼ in. (2.5 x 4.4 x .6 cm). Sterling silver, enamel. Photo by artist.

HOT TIPS
RIVETING

— Choose a drill bit to make a hole exactly the size of the rivet tube. The tube should fit tightly in the hole!

— To create perfectly aligned holes in two metal elements, drill the holes in the front piece first. Only drill holes in the back piece once you're finished with all enameling and firing. Metal expands and contracts in the kiln. If you drill rivet holes prior to enameling, you may find they don't match once you're finished enameling and ready to rivet.

— After enameling, carefully line up the two elements to be riveted. Only mark one hole for the back piece. Drill the marked hole first, insert a temporary piece of tube through it, and then mark and drill the second hole. Once two temporary tubes are in place, drill the rest of the holes.

— If you've soldered tube rivets to the back of the piece you're enameling so they won't show on the front, you can use a simple foam meat tray template to help mark the correct locations for drilling the back piece. Push the soldered rivets into the foam to mark it. Carefully double-check the rivet marks in the tray against the front piece. Place the tray template on the back metal piece, and drill through the marked points.

— File the tube rivet to the correct length. The amount to be flared should equal approximately half the tube's outside diameter. Extend this much tubing past each surface to be joined.

— Place a scribe or a small sharp tool into the tube and move it in a circular motion to flare the tube over the drilled hole, securing it in place.

— Flare the other end of the tube if needed. (If you soldered a decorative head to the other tubing end, flaring that end would not be necessary.)

— Prior to enameling, I most often solder the tube to the back of the metal with eutectic, IT, or hard solder. If you're careful not to over-fire, the soldered tube should hold up fine through multiple firings. If you're afraid of reflowing the solder joint, prior to firing you can paint it with a protective coating of white typing correction fluid or yellow ochre.

FABRICATED PRONG SETTINGS

Mickey Johnston, a fine jeweler and enamelist, created the following settings and prepared these samples to illustrate his techniques.

Simple Bezel Setting

1. Cut a piece of fine silver or fine gold bezel wire and form it to fit around the enamel to be set. Join the wire ends with hard solder.

2. Use medium solder to attach the bezel wire frame to a metal back plate.

3. Trim the outside edge of the metal back plate. Drop the enamel into the setting, and carefully push the bezel wire over it.

4. Burnish the bezel setting smooth around the enamel.

Pop-up Prongs

1. Trace the completed enamel onto a piece of 22- or 20-gauge metal. Draw a second line a few millimeters inside the first tracing, and draw areas to cut out as prongs.

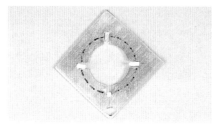

2. Use a jeweler's saw to cut out the interior shape, cutting around the prongs so they can be lifted. From the front of the metal shape, use pliers to lift the prongs.

3. Place the enamel on the "seat" you've cut, and press the prongs over the enamel to hold it in place.

This fancy version shows the back plate sawn and pierced into a decorative design.

Suspended Pronged Object

This method works especially well if you want to set two domed pieces against each other. Both sides can be equally visible and beautiful.

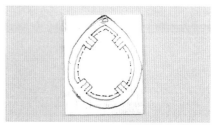

1. Trace the outside edge of the enamel onto a metal sheet. Scribe around the outside of the outline, leaving sufficient space for wide prong tabs.

2. Pierce out the center form. Make two vertical cuts in each prong tab.

3. Push down the center tab of each cut prong. Push the other two tabs up.

4. Drop the enamel into the setting, resting it on the center prong tab. Push the top two tabs over to secure the enamel.

Tab Frame Setting

1. Cut a strip of 20-gauge metal sheet to create a frame. Form it to fit tightly around the enamel you're setting. The rim of the strip should be taller than the enamel. Form a round wire to fit the inside edge of the strip.

2. Solder together the formed metal strip and the wire. (Alternately, you can solder the strip to a flat metal sheet to create a backing.) Solder a jump ring to the top of the strip if desired. Use a jeweler's saw to cut down three tabs from the metal strip to act as prongs.

3. Round the cut metal edges with a file. You may also want to file the tabs thinner.

4. Drop the enamel into the frame, and then push the prongs down to hold it in place.

Bead Frame Setting

1. Following the instructions for the tab frame setting, cut a strip of metal and form a frame that properly fits the enamel.

2. Create three identical fine silver balls and a jump ring to solder to the frame. (To make the balls, use a torch to heat short pieces of fine silver cloisonné wire until they bead up. For further instructions, see page 108.)

3. Solder the fine silver balls to the interior top edge of the frame. Thoroughly clean the soldered metal.

4. Using a jeweler's saw, cut prongs under each bead. Trim the cut prongs with a file and sand them smooth.

5. Drop the enamel into the setting with its front side against the beads. Push the prongs over the back side of the enamel.

6. If desired, you can drop a pierced or solid metal back plate into the setting before pushing over the prongs.

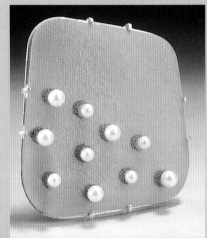

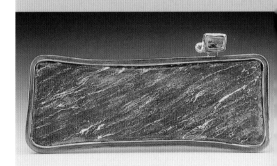

Top: Kimberly Keyworth *Tagetes Target Neckpiece*, 2003. 2¼ x 1¼ in. (5.7 x 3.2 cm). Sterling silver, 22-karat gold, enamel; fabricated, torch fired. Photo by George Post.

Center: Sun Hwe Park *Shadow B Brooch*, 2000. 2 x 1½ in. (5 x 3.8 cm). Sterling silver, pearls, enamel. Photo by Mark Johnston.

Bottom: Yoshiko Yamamoto *Brooch*, 1998. 4 x 2 x ¼ in. (10.2 x 5 x .25 cm). Copper, sterling silver, 22-karat gold, 18-karat gold, enamel, boulder opal, diamond. Photo by Dean Powell.

Frame Front, Tab-Back Setting

1. Cut a metal strip slightly taller than the enamel to be set. Form the strip to fit around the enamel and solder it together with hard solder.

2. Use medium solder to connect the formed metal strip to a piece of 20- or 22-gauge sheet metal.

3. Pierce a shape out of the sheet metal that is slightly smaller than the frame. The edge you create will be the shape of the frame on the front of the setting.

4. Drop the enamel into the frame, with the front side against the frame edge.

5. Saw out tab prongs, and push them over the enamel to hold it in place. (In this example, a pierced silver sheet was dropped in the setting on the back side of the enamel before the prongs were pushed down. If this were a

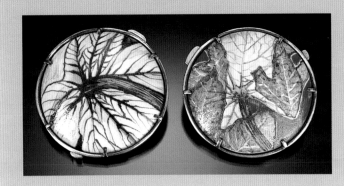

Mi Sook Hur
Brooches, 2003.
2 in. diameter (5 cm) each. Copper, sterling silver, enamel, watercolor enamel.
Photo by Robert Diamante.

brooch, you could solder a pin finding on the silver sheet.)

MOUNTING ENAMELED WALL PIECES

Select a frame or backing material that complements the finished enameled artwork. Plastic, metal, wood, slate, steel, cloth, and tile are just a few possibilities. An enamel can be glued into a frame or backing material, but it's a good idea to rough up the counter enamel before applying the adhesive.

Gluing & Grouting

Determine what type of adhesive to use based on whether the piece will be indoors or out and how it will function. For small pieces I recommend silicone rubber cement, which is similar to bathtub caulking. You can purchase this clear glue from hardware stores, from home improvement centers, or from stores that sell fish tanks and aquarium supplies. This adhesive has three major advantages: the silicone slightly cushions the enamel; the glue can be removed later should the enamel crack or require

repair; and if you apply an excessive amount, simply let it completely dry, and then cut and peel it away with a razor blade. (In addition to the silicone, I also try to have a mechanical device for holding the piece together, such as screws, rivets, or tabs.) Other artists recommend a hot glue that melts between 175°and 200° F (79.4° and 93.3° C). If an enamel piece ever needs to be removed, you can use a hot iron to re-melt this adhesive. Most commercial epoxies and super-strength glues are too brittle and don't hold up as well as a more resilient glue.

Some artists build a frame and then use tile grout to hold the enamel in place. To mount an enamel this way, mix the grout as directed, and when it begins to set, press the finished enamel into the grout. Be certain to wipe excess grout off the front of the enameled surfaces.

Using Backing Boards

If you drill holes into a metal piece prior to enameling, you can screw the finished artwork to a backing. Screwing together a wall-oriented enamel will make it easier to take it apart if needed. If you want the enamel to float off the backing surface, place tubing pieces over the screws to act as spacers. You also can use rivets to make a connection.

Enameling Techniques

William Harper *The Virgin and the Unicorn,* 1988. 6½ x 3 in. (16.5 x 7.6 cm). 14-karat gold, 18-karat gold, 24-karat gold, sterling silver, enamel, tourmalines, amethyst, pearls; cloisonné. Collection of The Victoria and Albert Museum, London, England. Photo by The Victioria and Albert Museum, London, England.

THIS SECTION IS WHERE you'll be able to apply all the fundamentals you learned earlier and start enameling! The techniques here are presented in the same order that I teach them to my university students. We begin with many ways to control sifted enamel, then learn to blend colors using drawing and painting techniques and liquid enamels. Practice firing, sifting, blending, and painting with enamel will help develop the enameling skills needed for pieces that are more labor intensive and time consuming in their metal preparation, such as cloisonné, champlevé, plique-à-jour, or dimensional pieces. Even though some students may feel impatient and want to jump in and make a champlevé or cloisonné piece in the beginning, they always thank me later for teaching them to enamel on pieces that do not involve long hours bending wires, etching, or sawing out sheet.

Browse through the various techniques presented in this section before you begin making your own samples. Choose those that appeal to you most, and read carefully to understand how they're made so that you can move ahead confidently.

If you're a beginner, I suggest you practice on inexpensive copper, making many samples as you improve your skill. Because of the rhythm of enameling procedures, (letting things cool, or dry) it's a good idea to work on several small sample pieces at once, trying different techniques and colors simultaneously, and always having another piece of copper ready for testing what you've learned on the last piece. For me, the only monotonous task is cutting and cleaning the metal and applying the base coats or counter enamels, so I do this on several pieces at one time, setting them aside for later use.

In most cases, for the sake of clarity, only one technique is used on pictured samples. It will be much more fun for you though, to try several techniques on the same sample piece. Keep the samples you make at your worktable and as you learn something new, pick one up and continue working on it, responding to what is already there and enjoying the new things you're learning.

Sifting Techniques

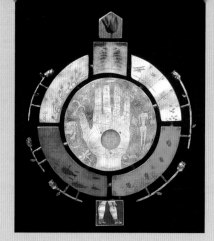

Right: Morgan Brig *Have You Ever Loved Your Life?*, 2003. 25 x 20 x 2 in. (63.5 x 50.8 x 5 cm). Copper, enamel, mixed media. Photo by Christopher Conrad.

Far right: Daniel Jocz and **Becky Brannon** *Marlene Series: Cherry Solitaire*, 1998. 2¼ x ⅞ x ⅞ in. (5.6 x 2.2 x 2.2 cm). Copper, silver, gold, enamel. Photo by Dean Powell.

THOUGH YOU MAY remember sifting as the enameling technique you learned in grade school or at summer camp, don't overlook it as a way to create exciting and expressive artwork. Sifting is also referred to as "dusting," and though the technique itself is simple, the effects that can be created are amazingly rich and varied. Transparent and opaque colors can be sifted and fired during the same firing. They can be used over and under each other, next to each other, overlapping or separated with soft or sharp edges. An easy way to learn about how your colors look together is to make several sifted pieces using colors in different combinations.

You learned how to sift evenly when applying base coats and counter enamels or made color tests. In this chapter, you'll learn other ways to control and work with sifting techniques to create special effects and interesting surfaces. Many thin layers of color can be sifted and combined for special effects, to create rich textured surfaces, and to define and draw shapes, lines, and patterns. With controlled firing techniques, the visual and tactile surface of even one single layer of sifted glass can be an effective way to define and emphasize form on dimensional work.

SIFTING OVER STENCILS

Stenciling is a simple way to create a design in enamel and to control the application of more than one color. A stencil is something you hold over the surface of the metal or enamel to block the grains of glass as they fall through the sifter. Make or use found materials for stencils and work in thin coats, with as many layers and firings as you like. Be creative as you choose stencils. You might try wet or dry brown paper towels, perforated or cut sheets of typing paper, metal that has been drilled, sawn, or etched completely through its surface, silk screens, open-weave fabrics, and organic or manmade shapes you live with every day. To capture fine details, sift through or around the stencils with 200-mesh or 100-mesh enamel so the smaller particles of glass will define the intricate areas of the design. With larger stencils, regular 80-mesh enamel works fine. You can sift more than one color on a piece for a single firing if you apply it thinly. If you apply too much enamel, it could build up and create expansion and contraction problems, resulting in cracks in the glass.

Don't forget to incorporate firescale into the design if you wish. If you leave some areas of the copper bare, they will be covered with firescale after firing. You can clean them in pickle or leave the areas dark as part of the design. If you later coat the firescale areas with clear enamel, they will appear reddish brown.

You can vary the effects of the stenciled design by working with opaque and also transparent color. The transparent color will allow the shapes beneath to show through, and the opaque color can be used to completely change or cover an area beneath it. Stenciling is a great first project if you are learning to enamel because you can quickly create designs and learn how colors work together while also getting plenty of practice firing.

Using Wet or Dry Paper Stencils

1. Cut stencil shapes out of soft paper. (I suggest brown paper towels because they are relatively lint

Sifting techniques may have been put into practice when enamellers first realized that no metal partitions were necessary in their work. At some point, someone decided it was easier to sift color across a large surface than to use a paintbrush. Covering an entire metal surface with enamel was first practiced by enamellists in Limoges, France, during the 15th century, but we don't know if they wet packed or sifted their base coat enamel layers.

Kohler Co. Enamel Shop, 1973. Courtesy Kohler Company, Kohler, Wisconsin

furnaces are 8 feet (2.4 m) deep, 6 feet (1.8 m) wide, and 6 feet (1.8 m) high. The cast iron pieces are loaded into the furnace and preheated to approximately 1600° F (871° C) before being removed. A sieve with an air vibrator is used to sprinkle enamel on the pieces while they are red hot. The pieces are then reloaded into the kiln where the enamel fuses to the cast iron. If the enamel was applied to cold metal, the glass would flow and fuse faster than the iron heats, causing crawling and faulty bonding.

In 1764, Robert Dossie wrote *The Handmade to the Arts*, an important publication regarding early craft practices. It appears from his text that prior to 1764 the technique of sifting was used for two enameling purposes: to classify the grain size, and to apply the enamel. Dossie explains in his text that the method was to, "rub the surface to be enameled over with oil of spike; and then, being laid on a sheet of paper or piece of leather, to save that part of the enamel which does not fall on a proper object, to sift on the oiled surface until it is the proper thickness." He explains that, "great care must be taken when using this method not to shake or move too forcibly the pieces of work thus covered with the powdered enamel."

In 1873 at the age of 29, John Michael Kohler, an Austrian immigrant, purchased an iron and steel foundry and began making steel implements for farmers, as well as castings and ornamental iron pieces. In 1883 he was launched into the plumbing business when he got the idea to bake enamel on a hog scalder, creating Kohler's first bathtub. It's interesting to witness the large-scale sifting and enameling currently being practiced at the Kohler Company. In their enamel shop, 33 enameling furnaces are presently in use. The interior dimensions of these

Sifting is a technique used by many enamellists in a variety of ways. Kenneth Bates, fondly known to many as the "Dean of American Enameling" is no exception. A master of many enameling techniques, he began studying enameling in 1924 and only a few years later began teaching it to his students at the Cleveland Institute of Art. Bates' influence as a teacher and author of enameling books has been significant and many important American enamellists working from the 1930s, through the 1960s and 1970s learned their craft from Kenneth Bates.

Kenneth Bates *Plaque with Trees*, 1993. 9 x 9¼ in. (22.9 x 23.5 cm). Copper, enamel. Photo by Bernard Jazzar.

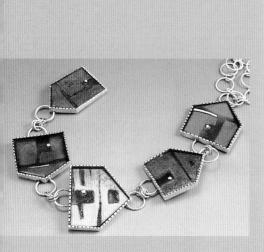

free, but other papers or cloth may
also work well.)

2. If you want to create a design
with hard, crisp edges, soak the
stencil in a dish of water. Squeeze
out the excess water, and apply the
wet stencil to the clean copper or
enameled surface as shown.

3. Thinly and evenly sift the powdered enamel color of your choice
over the stencil. Pick up the stencil
with tweezers (see photo), or let it
completely dry before removing it.
If the stencil is too wet or otherwise
difficult to lift off the piece, sew a
stitch of thread through it so you
can lift the thread and easily
remove the stencil.

4. Let the enamel completely dry,
and then fire the piece.

5. If you want the next stenciled
layer to have softer edges, sift the
enamel over a dry paper towel.

6. If you want an even softer edge,
raise the stencil off the piece when
you sift a new enamel layer.

The samples below by Mi Sook Hur
show the effects of sifting in layers
using a paper stencil with transparent and opaque enamel colors.

HISTORICAL HIGHLIGHT

Sifting techniques may have been put into practice when enamellers first realized that no metal partitions were necessary in their work. At some point, someone decided it was easier to sift color across a large surface than to use a paintbrush. Covering an entire metal surface with enamel was first practiced by enamellists in Limoges, France, during the 15th century, but we don't know if they wet packed or sifted their base coat enamel layers.

Kohler Co. Enamel Shop, 1973. Courtesy Kohler Company, Kohler, Wisconsin

furnaces are 8 feet (2.4 m) deep, 6 feet (1.8 m) wide, and 6 feet (1.8 m) high. The cast iron pieces are loaded into the furnace and preheated to approximately 1600° F (871° C) before being removed. A sieve with an air vibrator is used to sprinkle enamel on the pieces while they are red hot. The pieces are then reloaded into the kiln where the enamel fuses to the cast iron. If the enamel was applied to cold metal, the glass would flow and fuse faster than the iron heats, causing crawling and faulty bonding.

In 1764, Robert Dossie wrote *The Handmade to the Arts*, an important publication regarding early craft practices. It appears from his text that prior to 1764 the technique of sifting was used for two enameling purposes: to classify the grain size, and to apply the enamel. Dossie explains in his text that the method was to, "rub the surface to be enameled over with oil of spike; and then, being laid on a sheet of paper or piece of leather, to save that part of the enamel which does not fall on a proper object, to sift on the oiled surface until it is the proper thickness." He explains that, "great care must be taken when using this method not to shake or move too forcibly the pieces of work thus covered with the powdered enamel."

In 1873 at the age of 29, John Michael Kohler, an Austrian immigrant, purchased an iron and steel foundry and began making steel implements for farmers, as well as castings and ornamental iron pieces. In 1883 he was launched into the plumbing business when he got the idea to bake enamel on a hog scalder, creating Kohler's first bathtub. It's interesting to witness the large-scale sifting and enameling currently being practiced at the Kohler Company. In their enamel shop, 33 enameling furnaces are presently in use. The interior dimensions of these

Sifting is a technique used by many enamellists in a variety of ways. Kenneth Bates, fondly known to many as the "Dean of American Enameling" is no exception. A master of many enameling techniques, he began studying enameling in 1924 and only a few years later began teaching it to his students at the Cleveland Institute of Art. Bates' influence as a teacher and author of enameling books has been significant and many important American enamellists working from the 1930s, through the 1960s and 1970s learned their craft from Kenneth Bates.

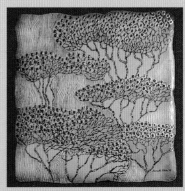

Kenneth Bates *Plaque with Trees*, 1993. 9 x 9¼ in. (22.9 x 23.5 cm). Copper, enamel. Photo by Bernard Jazzar.

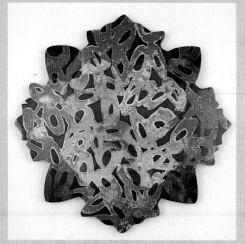

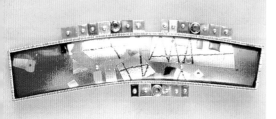

free, but other papers or cloth may also work well.)

2. If you want to create a design with hard, crisp edges, soak the stencil in a dish of water. Squeeze out the excess water, and apply the wet stencil to the clean copper or enameled surface as shown.

3. Thinly and evenly sift the powdered enamel color of your choice over the stencil. Pick up the stencil with tweezers (see photo), or let it completely dry before removing it. If the stencil is too wet or otherwise difficult to lift off the piece, sew a stitch of thread through it so you can lift the thread and easily remove the stencil.

4. Let the enamel completely dry, and then fire the piece.

5. If you want the next stenciled layer to have softer edges, sift the enamel over a dry paper towel.

6. If you want an even softer edge, raise the stencil off the piece when you sift a new enamel layer.

The samples below by Mi Sook Hur show the effects of sifting in layers using a paper stencil with transparent and opaque enamel colors.

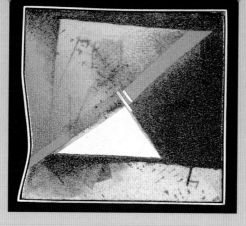

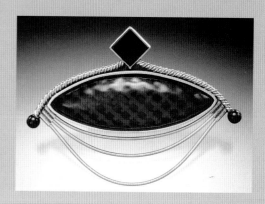

Right: Rebekah Laskin
Brooch, 1987. 2⅛ x 2⅜ in. (5.4 x 6 cm). Enamel, copper, sterling silver. Photo by artist.

Far right: Barbara Minor *Red & Black Marquis Pin*, 2001. 4 x 2 in. (10.2 x 5 cm). Copper, silver, fine silver, onyx, enamel. Photo by Ralph Gabriner.

Patterned Stencils

In addition to creating shapes and textures, you can create patterns with found or handmade stencils. Barbara Minor created the following samples using only one color to show how a stencil creates pattern. On a finished piece, the stencils would be used with multiple transparent or opaque colors in numerous firings. The finished pieces could include patterns that can be seen through transparent colors, with layers of other patterns which overlap the design beneath them.

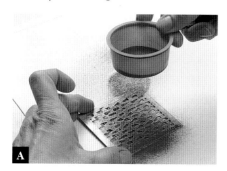

This stencil is a copper sheet with etched perforations (photo A). 100-mesh red enamel was sifted

through it onto a base coat of white enamel (photo B).

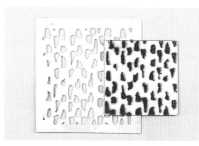

This is the patterned stencil piece after one firing.

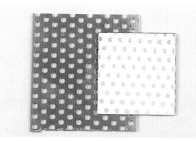

Another idea for a patterned stencil is a found piece of perforated brass screen. The color was sifted onto a base coat of white enamel.

Patterned Stencils on Copper with Firescale

If you sift enamel through a stencil onto bare copper you can create firescale patterns. Once you fire the piece, you can sift more layers of enamel through the same stencil, firing between siftings. If you use an enamel with a high acid resistance (most lead-free colors) for at least the first sifting, you can clean parts of the

piece in pickle to control the amount of firescale left on the piece before the final transparent overcoat. To produce the reddish copper color shown in the samples below, leave more firescale beneath the transparent flux. To produce the golden clear color, fire the clear enamel very hot over very clean copper.

These test pieces by Jessica Turrell investigate the effects of firescale used with sifting techniques.

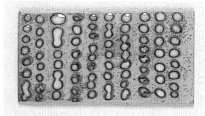

For this sample, clear enamel (flux) was sifted through a stencil made of metal with drilled holes and fired. The metal was cleaned in a pickle bath.

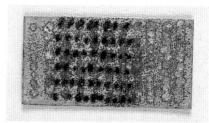

This sample shows black enamel sifted onto the piece through the same drilled metal stencil and fired. The firescale was left on the surface.

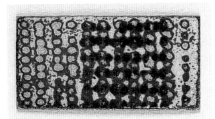

Clear enamel (flux) was sifted over the firescaled surface and fired.

The piece was fired at a higher temperature so the copper oxide would affect the color of the clear enamel.

Stenciled or Painted Pattern Using Firescale with Sifting

The following samples show Jessica Turrell's continued experiments with patterning and firescale. Ball clay or Scalex can be used for the technique.

Ball clay was painted on the copper to act as a resist and allowed to completely dry. The piece was heated to create a firescaled surface on the exposed copper. Clear enamel was sifted over the entire surface (the ball clay and the firescale) and the piece was fired. After firing, the enamel could be removed from the places where the ball clay was applied, revealing clean copper. Clear enamel was sifted over the entire surface. When fired, the clear enamel appears bright gold over the clean copper and reddish brown over the firescale.

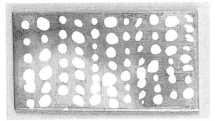

Paint or stencil ball clay onto a clean copper surface, and let dry.

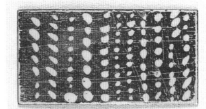

Fire the piece in the kiln until firescale appears on the uncoated copper. After removing the piece from the kiln, apply clear enamel over the entire surface of the metal and fire it again. Flake off the ball clay resist and the clear enamel covering it to reveal areas of clean copper. Rinse the metal. Sift clear enamel over the entire surface and fire until fused. The areas once covered with ball clay will be shiny and bright. The areas coated with firescale will appear reddish brown.

SIFTING & DUMPING

To begin this technique, you'll need a piece of metal with a clean surface or a fired basecoat. You apply a holding agent or adhesive to the metal, sift on a coat of dry enamel, and tap off the excess powder. Once the piece is completely dry, fire it. Experiment with this technique: consider forming fine lines, large shapes, dots, patterns, and even lettering. In addition to common enamel holding agents, consider using glycerin, oil binders, stamping ink, or fiber-tipped marking pens to hold the enamel in

place. Some brands of pens are more suitable for this task than others, so experiment to determine which ones stay wettest and hold the powder best.

Depending on the degree of detail in your design, 200-mesh enamel may be the best size to use. (Remember that finer glass particles cling more easily to the detailed lines.) Prepare the enamel and have it in the sifter before you apply the holding agent to the piece. You need to be able to quickly sift the enamel over the design before it dries.

Created by Sarah Perkins, the following samples of the sifting and dumping technique illustrate some of the detailed lines and patterns you can make.

This sample shows the variety of line widths you can create using different binders. An opaque ivory base coat was fired on the metal. Lines were drawn on the base coat with different size felt-tip markers, with gum binder on a paintbrush, with water on a paintbrush, and with thinning oil on a crow quill pen. 200-mesh opaque red enamel was sifted on top of the lines, the excess powder was tapped off, and the piece was fired until fused.

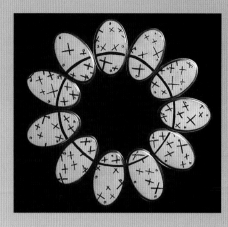

Right: Gretchen Goss *Coreopsis*, 2000. 16 x 20 in. (40.6 x 50.8 cm). Copper, enamel. Photo by artist.

Far right: Sarah Turner *All For One Buckles*, 2002. 3 x 2 in. (7.6 x 5 cm). Copper, sterling silver, enamel. Photo by Courtney Frisse.

In this sample, an opaque ivory base coat was fired on the metal. Hand cream was rubbed onto a fingertip. The greased finger was printed onto the base coat. 200-mesh opaque blue enamel was sifted over the fingerprint, and the excess powder was tapped off. The piece was fired until fused.

For this sample, a rubber stamp was used with embossing fluid to make a pattern over an opaque black base coat. 200-mesh

opaque blue enamel was sifted over the pattern, the excess powder was tapped off, and the piece was fired. Embossing fluid was stamped again in a different place on the piece, and 200-mesh opaque yellow enamel was sifted on the pattern. The excess powder was tapped off before the piece was re-fired.

To create the samples below, Elizabeth Turrell used a hand-cut stamp with ink from a regular office stamp pad. Immediately after stamping the ink, 200-mesh enamel was sifted over the whole piece. The piece was picked up, tapped sharply to remove the excess enamel, and dried for a few minutes before firing.

Here's the sample with opaque black enamel sifted over a red base coat, and fired. The sample above right used the same technique, but opaque white enamel

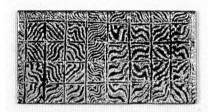

was sifted over a black base coat, and then the piece was fired.

A hand-cut stamp was used to make the pattern.

SGRAFFITO THROUGH SIFTED ENAMEL

Sgraffito is an Italian word that describes the technique of scratching through a layer of something to expose what is beneath it. Potters scratch through glazes or slips to expose clay, and painters use sgraffito to expose other layers of paint. On page 99 you'll learn about sgraffito with liquid enamel, an especially nice technique that can result in very fine and delicate scratch lines. When combined with stenciling or other

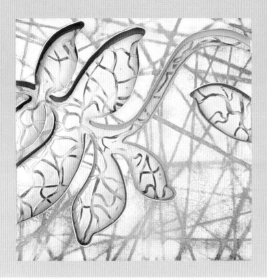

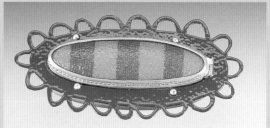

Left: Sarah Krisher *Complicating the Simplified Series,* 2003–2004. 12 x 12 in. (30.5 x 30.5 cm). Enamel, copper. Photo by Robert Muller.

Center: Jan Baum *100% Pure,* 1999. 2¾ x 1⅜ x ¹³⁄₁₆ in. (6.9 x 3.5 x 2 cm). Sterling silver, copper, enamel, found object.

Right: Julie Brooks Price *Two of a Kind-Male,* 2001. 1¼ in. (3.2 cm). Silver, copper, enamel. Photo by Linda Darty.

sifting techniques, however, sgraffito can be very effective for creating textured surfaces. You can scratch a design through any enamel color to expose either the metal beneath it or a previously fired enamel layer. To scratch through unfired enamel, you can use many tools, such as needles, toothpicks, or any other sharp pointed instruments. For removing larger areas of color, you can use paintbrushes, erasers, or even your fingers.

How to Sgraffito Sifted Enamel
Before You Begin
• Position a clean piece of paper beneath the work to catch extra enamel so you can reuse it.

• Wear a dust mask so you won't inhale any enamel powder while you're scratching it away.

• If your design is going to be very precise, with fine lines, prepare 100-mesh or 200-mesh enamel.

1. If the piece you want to sift requires the use of a holding agent, apply it with a paintbrush or sprayer.

2. Sift the enamel onto the metal. If used, let the holding agent dry.

3. Scratch away lines or areas as desired. Blow or brush the excess enamel onto the clean catch paper.

4. Fire the piece until the enamel fuses.

In the photo to the right, sample A features a fired bright blue opaque base coat.

On sample B, a second coat of lime green enamel was sifted over the bright blue. A sharp tool was used to scratch a line through the lime green enamel to reveal the blue base coat.

Sample C shows the same sgraffito process executed in reverse. Bright blue enamel was sifted over lime green, and then scratched through to reveal the base coat.

A

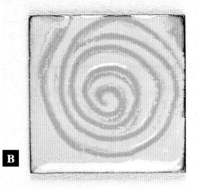

B

C

SGRAFFITO THROUGH FIRED ENAMEL

Kimberly Keyworth has developed a unique method for using sgraffito after a piece has been fired. This method creates particularly sharp and clear edges that can enhance a design.

This sample shows three coats of opaque enamel color sifted over a black base coat, fired between layers.

In this sample, the interior design has been ground out with diamond burrs, and then refilled with 325-mesh wet enamel.

Here, diamond burrs have been used to grind out another design from the enamel.

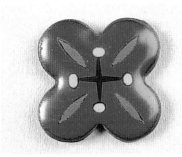

325-mesh enamel was inlaid into the ground-out area, and this is the finished sample after it's been fired and matte finished.

SIMPLE SIFTING

I use the term "simple sifting" to refer to the direct process of using enamel as pure color on a metal form. Depending on the color chosen and the way it's fired, a single coat of sifted enamel can be a provocative addition to a cut or formed metal shape. There are many ways you can create metal forms for simple sifting, such as raising, fold forming, electroforming, die pressing, and casting. To convey the simplicity of enameling intricate and dimensional shapes, here are two uncomplicated ways you can make pierced dimensional pieces using only sawing skills and minimal forming.

Cut & Peel with Simple Sifting

You can effectively use paper folding and cutting techniques to create interesting metal forms for simple sifting. You can use this technique on any gauge metal that you can peel up with pliers. Depending on the size and the function of the piece, I generally use 26- to 18-gauge metal. Since 100-mesh enamel is less likely to bounce off the edges of tiny cut metal shapes, you should use graded sifters to remove the larger 80-mesh particles from the powder prior to sifting.

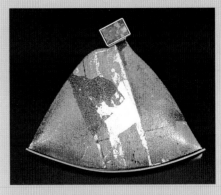

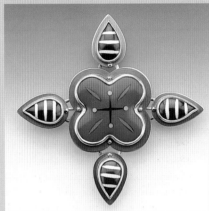

Top: Rebekah Laskin Untitled, circa 1985. 2 x 2 in. (5 x 5 cm). Copper, enamel, 18-karat gold, 24-karat gold foil, lapis lazuli.

Center: Kimberly Keyworth *Red Fire Poppy Brooch*, 2003. 2¼ x 2¼ in. (5.7 x 5.7 cm). Sterling silver, 22-karat gold, enamel; fabricated, torch fired. Photo by George Post.

Bottom: Lenore Davis *Dancers Playing Baseball*, 1983. 9 in. diameter (22.9 cm) Enamel, steel flanged tile. Photo by Keith Wright.

In the brooch above I pierced a piece of copper with a jeweler's saw, and the sawn sections were peeled up with pliers. The form was then coated with one coat of enamel and fired. Coating the cut and peeled piece with an adequate amount of holding agent allowed the sifted enamel to properly adhere to the metal. You can achieve variations in a single coat of color by controlling the thickness of the enamel and the time of the firing.

1. Draw a design on paper that includes decorative shapes to saw out and peel up. You can even practice this cutting and peeling technique on the paper model of your design. Glue the drawing to the metal, or re-draw the design directly on the metal with a fine-tip permanent marker. Drill a small hole through the metal at the end of each saw line as shown.

2. Thread the saw blade through one drilled hole, and saw the line. Remove the blade, reinsert it into the next hole, and saw another line. Repeat this process to saw all the lines in your design.

3. Using a plastic or rawhide mallet, form the sawed metal piece in a wood block or a stump to give it structural strength and lessen the possibility of it warping after firing.

4. From the back side of the metal piece, use pliers or a pointed tool to push out the sawed shapes as shown. If needed, file the cut metal edges smooth.

5. Carefully paint or spray a holding agent onto all areas of the metal surface.

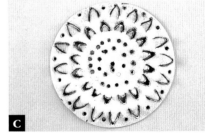

6. While the binder is wet, sift 80-mesh to 150-mesh enamel onto the piece. Because the firing trivet would mark the domed metal surface, first sift on a counter enamel and fire the piece (photo A). Next, sift on a base coat (photo B) and fire the piece until the enamel is fused (photo C).

Left: **Linda Darty** *Outside In: Vase*, 2002. 6 x 5 x 4 in. (15.2 x 12.7 x 10.2 cm). Copper, sterling silver, enamel. Photo by artist.

Right: **Renee Menard** *Landscape I*, 2002. 11 x 14 in. (27.9 x 35.6 cm). Copper, enamel. Photo by Helen Shirk.

Far right: **Linda Darty** *Outside In: Cup*, 2002. 6 x 5 x 4 in. (15.2 x 12.7 x 10.2 cm). Copper, sterling silver, enamel. Photo by artist.

Pierced Piece Using Simple Sifting

The simple sifting technique is also very effective on pierced copper. When designing a piece, keep the pierced sections small and connected with varying line widths of metal. Without support, the enamel is more likely to chip off very fine lines and very long lines. Carefully draw your design directly on a metal piece, draw it on tracing paper and adhere it to the metal with rubber cement, or use any other transfer method you prefer. (When using rubber cement, apply it to both the metal and the paper, let it dry on both surfaces until tacky, and then adhere the two pieces. If you don't use this method, your drawing could slip while you're sawing out fine lines.) When sawing fine lines on metal you wish to enamel, select metal that is at least 22- gauge thick. (Enamel is likely to crack or chip off thinner sawn metal.) It's easiest to saw very fine lines if you use a size 4/0 or 5/0 blade. Since 100-mesh or 200-mesh enamel will more easily adhere to fine lines, use graded sifters to remove larger particles of enamel. You'll later use this same piercing technique to create champlevé and plique-à-jour pieces.

A

B

1. Drill a hole in the metal for each pierced area of your design. Insert the saw blade through one hole and saw out the interior shape (photo A). Repeat this process to saw out all shapes (photo B).

2. Form the pierced metal piece to give it strength, lessen the possibility of it warping, and decrease the chance of the enamel cracking or chipping off the surface.

3. To help prevent the delicate piece from warping, you'll fire both the base coat and the counter enamel on the pierced metal at the same time. Paint a holding agent on the back side of the clean metal and sift on the counter enamel. Lightly spray a holding agent over the counter enamel. Holding the piece by its edges, turn it over to work on the front side.

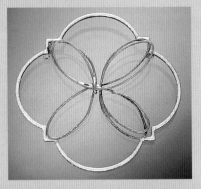

Esther Knobel Display from the series *My Grandmother is Knitting Too*, 2002. 6¾ in. (17.1 cm). Copper, enamel; knitted. Photo by Vered Kaminski.

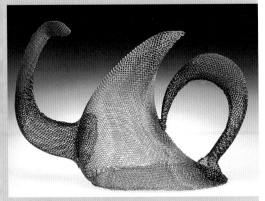

Top: **Masumi Kataoka** *Where It Ages Like Tree Rings #1*, 2002. 10 x 8 x ½ in. (25.4 x 20.3 x 1.3 cm). Sterling silver, copper, enamel, nylon, wood; fabricated, electro-formed. Photo by artist.

Center: **Jan Baum** *Fountain Flower #2*, 2003. 2½ x 2½ x 5⁄16 in. (6.4 x 6.4 x .8 cm). Steel, enamel. Photo by Norman Watkins.

Center: **Mary Chuduk** *Tea Strainer Teapot*, 2002. 8 x 12 x 6 in. (20.3 x 30.5 x 15.2 cm). Copper screen, liquid enamel; sprayed, formed, kiln fired. Photo by Jeff Scovil.

Bottom: **Natalya Pinchuk** *Bracelets*, 2003. 4½ x 4½ x 2 in. (11.4 x 11.4 x 5 cm) each. Copper, silver plate, enamel. Photo by artist.

C

D

4. Spray a holding agent on the front side of the pierced piece, and then sift on a base coat of 100-mesh or 200-mesh enamel (photo C). Place the metal on the firing trivet with the top side facing up, and allow it to thoroughly dry. Fire the pierced and sifted piece until the enamel fuses. A second coat of enamel was sifted on the fused base coat and under-fired to "sugar" coat the surface (photo D).

SIMPLE SIFTING ON WIRE FORMS

In addition to simply balling up wire, you can twine it, knit it, weave it, crochet it, and use many other exciting textile techniques. You can create wire projects to enamel in copper, steel, or fine silver. It's simple to enamel wire forms. A properly applied holding agent will help dry enamel adhere to the wire. Because smaller grains more easily cover wire shapes, it's most effective to use 100-mesh or 200-mesh enamel. You also can use liquid enamel to create interesting tonal effects in the high and low areas of wire structures. (See page 97 for more information on liquid enamel.)

The original wire form

Spraying a holding agent on the wire form

Sifting enamel on one side of the wire form

Spraying holding agent over the sifted enamel

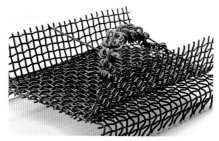

Sifting enamel on the reverse side of the wire form

The sifted and fired wire form supported on a screen

The enameled wire form

— Always wear a high-quality dust mask when sifting so you won't inhale the airborne glass particles that are invisible to your eyes. If you plan to sift a lot, you may want to purchase an air purifier and place it on a table near your sifting area.

— Sift enamels at a 90-degree angle to the surface of a piece. If needed, adjust the angle of the metal, not the angle of the sifter.

— Clean sifters by sharply tapping them against a table edge. Do this each time you change enamel colors.

— Clean very fine mesh sifters with a pressurized air sprayer (commonly used for cleaning computer keyboards). You can purchase one at an office supply store.

— Sift over two clean magazine pages or copy paper sheets. (It's easier to pour enamel off a slick,

glossy paper than newsprint.) The bottom layer keeps the back side of the top sheet clean. When you use the top sheet as a funnel to return excess enamel to its container, no dirt will fall into the container and contaminate the enamel.

— To lessen enamel dust in your studio, fold used catch papers twice before disposing them in a waste receptacle.

— When sifting several enamel colors on one piece for a single firing, arrange different sifting stations and assign a different color and sifter to each station. As you work, walk to the different stations to sift different colors onto the piece. If you follow this procedure, you won't have to put the piece down, and you won't contaminate the colors.

— Use 100-mesh to 200-mesh enamel when sifting intricate shapes or thin lines. The smaller grains of glass more easily attach to the fine detail.

— Use 80-mesh or larger enamel particles if you want transparent colors to be very bright and clear.

— Apply a holding agent to three-dimensional shapes before sifting. If needed, you can spray on the binder after sifting to "glue" the enamel grains in place. This top coat allows you to turn a piece over and sift the other side.

— Frequently damp wipe the surfaces in your studio!

Basse Taille

Jane Short *Ode to Joy*, 2002.
9 in. (22.9 cm). Silver, enamel;
champlevé, basse taille. Photo
by Clarissa Bruce.

NOW THAT YOU CAN SIFT, you
should try it with beautiful transparent colors over textured metal. The
effects are stunning! The transparent
glass is truly at its best when light
passes through it to reflective metal
surfaces. The French word, *basse
taille*, means "low cut" and though
we know the technique has been
practiced since the 14th century, the
term began to be used in the 16th
century to describe enamel pieces
that had mostly chiseled or engraved
surfaces beneath transparent glass.
Though the surface of the glass is
smooth, in the lower areas of the textured design, where it's thicker, it
appears darker. On higher metal surfaces it's thinner and therefore lighter
in value. Contemporary enamelists
use a variety of techniques to create
textured surfaces in metal, and then
either sift or wet inlay with clean,
washed transparent color. Basse taille
enameling is frequently combined
with other techniques such as champlevé, limoges, or cloisonné.

In addition to creating interesting
surfaces that can be seen through
the glass, many of the techniques
described in this chapter can also
be used to create even deeper
recessed areas that can be inlayed
with enamel for champlevé techniques (see page 116). The difference is that basse taille enameling

refers to transparent enamel covering both the low and high cut areas
in the design, while in champlevé
enameling, though the same metalworking techniques might be used
to create the recesses, only the low
areas are enameled, leaving the
high spots as bare metal. A piece
such as Jane Short's, in the photograph above, would be described
as basse taille and champlevé
because she hand-engraved the
metal to create both the textured
basse taille surfaces, and also the
recesses for the champlevé enamel.

PREPARING METAL FOR BASSE TAILLE
Prior to basse taille enameling,
you'll need to sand and finish, and
perhaps anneal the metal on which
you'll be working.

Sanding
Sand and finish metal before applying basse taille textures. The metal
should be bright and reflective
underneath the glass without
scratches or sanding marks. (If you
wait to sand out scratches until after
you texture the metal, you might
lose some of the texture.) Start by
using either a wet green kitchen
scrub pad or wet 320-grit carborundum paper to remove scratches from
the metal. Next, sand the metal with
wet 400-grit carborundum paper,

and in some cases, continue sanding
with 600-grit paper or steel wool.

Annealing
For some techniques it's necessary
to soften, or *anneal*, the metal
before you texture it. To accomplish
this, heat the piece until it's cherry
red, and then quench it in water or a
pickle solution. (If you want a visual
indicator to tell you how hot to
anneal the metal, paint a little soldering paste flux on the piece.
When it becomes clear and glassy,
the metal is annealed.)

ENAMELING THE METAL AFTER CREATING TEXTURE
After creating surface texture on the
metal, clean it as you would for all
enameling procedures. (Depending
on the texturing method used, I
might burnish the piece under running water with a very soft brass
brush and liquid soap.) Test the transparent colors you want to use on the
same base metal as the textured
piece. To achieve the greatest clarity,
be sure to sift the fine particles out of
the enamel or wash the enamel well.
Apply counter enamel to the back
side of the basse taille piece, and
apply clean, transparent colors to the
front of the piece. Depending on the
shape of the piece, you can sift or
wet inlay the transparent color. If
you're working on copper, fire the

HISTORICAL HIGHLIGHT

The guilloché style of basse taille enameling was developed during the industrial growth period of the late 1800s. Guilloché refers to a type of abstract geometric pattern cut into the surface of metal that you can see beneath transparent enamel. Once engraving lathes were developed, many metal pieces with a variety of machine-turned surfaces could be made. Wave designs, moiré patterns, and other motifs with fine lines could all be created with machine lathes. From the late 1890s and into the early 20th century, skilled engine turners made many enameled luxury items. Turning a perfect pattern

Attributed to the House of Peter Carl Fabergé (1846–1920), Notebook Cover, circa 1900. 4¼ x 2¹³⁄₁₆ in. (10.7 x 7.1 cm). Gold, enamel, diamonds. Bequest of Charles Phelps and Anna Sinton Taft. Taft Museum of Art, Cincinnati, Ohio.

over a shaped form was difficult for the workers because every mark on the metal surface showed beneath the transparent glass. For the most brilliant transparency, the finished glass surface needed to be very, very thin, and the coats of enamel were ground down smooth between each firing. Sometimes the final surface was even acid-etched to make certain all the grinding marks were erased before the last firing. This was done in an effort to make the final enameled surface perfectly glossy and smooth.

The Russian house of Fabergé used the guilloché technique for mass-producing a wide variety of luxury items in the late 19th century. After their pieces were exhibited at the Paris Exposition in 1900, the technique became more and more popular in workshops in Switzerland, England,

and France. These countries were soon in competition with Fabergé, making such items as cigarette cases, compacts, picture frames, clocks, and opera glasses for the wealthy to collect.

Fabergé is most well known for their impressive Imperial eggs, which were commissioned by the Czar Alexander III as a gift to his wife each Easter. The first egg was so well received that Alexander III ordered a new egg be made each year. This commission was continued by his son, who increased the order to two eggs each year, so he could give one to his mother and also to his wife. Given complete freedom to ignore cost and time, Fabergé enjoyed the challenge of creating something new and spectacular each year. Many of these eggs include guilloché basse taille surfaces.

Items such as these spoons and salt and pepper containers were made using the guilloché technique, probably during the first quarter of the 20th century.

Guilloché Salt and Pepper Shakers, early 20th century. 2½ x 1½ in. (6.4 x 3.2 cm). Sterling silver, enamel. Collection of Linda Darty. Photo by Linda Darty.

Guilloché Spoons, English, early 20th century. 4 x ¾ in. (10.2 x 1.9 cm) each. Sterling silver, enamel. Collection of Linda Darty. Photo by Linda Darty.

colors hot until they are bright and clear. (For further information about firing transparent colors on copper, see page 49.) Keep in mind that some transparent colors need a clear undercoat so they don't look muddy on both copper and silver. When choosing your clear enamel base coat, be sure to use the right type for the right metal. Copper flux will look yellow on silver. Keep a list as you discover which colors work well directly on silver or copper. This way, you can often refer back to your notes and know immediately when a clear undercoat is needed. (Warm colors usually need a clear base coat if they are to be fired on silver.)

APPLYING UNDERGLAZE PIGMENT TO TEXTURED METAL SURFACES

Most historical basse taille enamels do not include color in the recessed metal surfaces other than the one layer that coats the entire piece and creates subtle variations in value over its surface. To emphasize the textures, however, you may want to use black underglaze in the recessed areas.

1. Mix the underglaze with oil or water. If you use water-based binder before applying binder, the underglaze pigment might bleed out of the recesses.

2. Paint the underglaze on the metal. Wipe the pigment off the high points of the textured surface, and let it dry thoroughly.

3. Apply a holding agent if needed, and then sift on the transparent enamel. (Don't use a binder for the sifted coat of enamel if you don't have to—the enamel may be more transparent.)

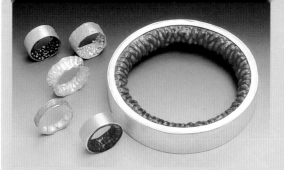

Anna Gallof *Sensual,* 2000. 1 in. diameter (2.5 cm), each ring; 1 x ¾ x 3½ in. (2.5 x 1.9 x 8.9 cm), bracelet. Sterling silver, copper, 24-karat gold foil, enamel; repoussé. Photo by Bob Barrett.

Right: David C. Freda *Stag Beetles, Grubs, and Raspberries,* 2001. 4½ x 2 x 1½ in. (11.4 x 5 x 3.8 cm). Fine silver, sterling silver, 24-karat yellow gold, 18-karat yellow gold, enamel. Photo by Barry Blau.

Right center: Sarah Letts *Brooch,* year TBD. 2¼ x 2⅞ in. (5.7 x 7.3 cm). Silver, enamel; champlevé.

Far right: Deborah Lozier *Bracelet Study in Blue,* 1998. 4 x 4 x ¾ in. (10.2 x 10.2 x 1.9 cm). Copper, enamel; fabricated. Photo by artist.

METAL-TEXTURING TECHNIQUES

There are many ways you can create interesting metal surfaces prior to basse taille enameling. Some of the same techniques are used to form pieces so they are enameled in high relief, or "en ronde basse."

Hammering

A

Metalworking hammers have different shapes and faces. By striking a piece of metal on top of a steel block or anvil, you can cause deep impressions and create a rich surface (photo A). To make bright marks, be sure the hammer face is well-polished and the steel beneath the piece is clean and smooth. You can even make interesting patterns on old hammer faces with a file or saw, and then transfer these textures into the metal via hammer blows.

Chasing & Repoussé

The chasing and repoussé technique involves pushing metal from

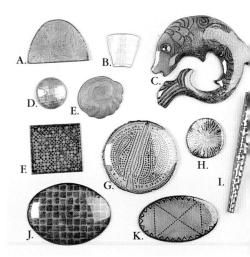

A. Hammering (Robert W. Ebendorf); B. Punching (Jesse Bert); C. Chasing (Robert W. Ebendorf); D. Roll milling (Jesse Bert); E. Repoussé (Whitney Boone); F. Etching (Linda Darty); G. Chasing (Suzanne Pugh); H. Hammering (Robert W. Ebendorf); I. Punching (Linda Darty); J. Chiseling (Jesse Bert); K. Stamping (Robert W. Ebendorf).

both sides with a variety of polished steel tools, like the ones shown in photo B. Chasing refers to pushing back metal from the front,

B

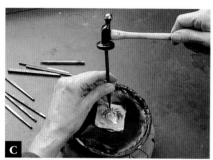

and *repoussé* describes pushing it out from the back. The word *repoussé* comes from a French term meaning "to push back." It's one of the oldest metalworking techniques, used for centuries to create dimensional surfaces for enameling, especially basse taille and champleve. During the process, metal is often held in pitch for support (see photo C). Use a bushy torch flame or a heat gun to slowly soften pitch. Take care not to burn or bubble the pitch, as this would cause it to be brittle and less resilient. Once you understand chasing and repoussé, you can adapt the process and use the same tools to create different textures, with or without using pitch as the backing.

1. Clean the metal surface.

2. Use a permanent marker to draw a design on the metal. Place the metal in the softened pitch.

3. Following the marked lines, use a liner tool to chase the drawing into the metal. Use tools to create dimension by pushing the metal from one side until the metal becomes work-hardened and may seem to lift slightly from the pitch.

4. Gently re-warm the piece, being careful not to burn the pitch, and lift the metal out of the pitch.

5. Clean the metal by using a heat gun or soft torch flame to melt off as much pitch as possible. To remove the greatest amount of residue, either wipe the warm metal with a paper towel, dissolve the pitch with turpentine, or burn the remaining pitch with a torch until it becomes white and disappears. Re-anneal the metal and clean it in a pickle solution.

6. Set the metal into the pitch so it can be worked from the opposite side.

7. Repeat this process many times until you achieve the desired surface dimension.

Line Tracing with Chasing Tools

You can use steel chasing tools and a hammer to make linear depressions in metal that you can later cover with transparent enamel. Depending on the type of line you want to create, you can place the metal in pitch or simply clamp it to a piece of steel or wood. Using the tool is similar to drawing on metal. Simply move it across the surface to create lines. Placing a drop of oil on the metal can make for smoother chasing, but is not always necessary. Hold the tools at an angle to the metal so you can use a pulling motion to easily guide them across the surface.

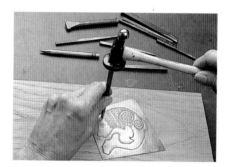

Line-chasing techniques were used to create this fish, and then the form was pushed out using repoussé.

Punching

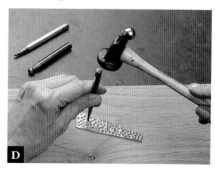

Punching is a very simple way to create a textured pattern on metal, and beneath transparent enamel, it shows up well from either side. To create dot-like textures on sheet metal, you can use a sharpened nail or a small chasing tool. If you punch the annealed metal on top of a piece of steel, it will remain flat with patterned marks. If you place the metal on wood or pitch before punching, the mark will be more dimensional (see page 77, photo D).

Stamping

You can purchase different commercial tools with which to stamp metal as shown above or make your own out of hardened steel rod. Anneal and clean the metal you want to stamp, and then position it on a smooth steel block or anvil. Hold the tool on the metal

and strike it hard with a hammer to stamp the metal surface (photo A). This is the same technique some metalsmiths use to hallmark their artwork with their name or to indicate metal type.

Chiseling

Chisels are hard sharp steel tools with different shaped heads (see photo above). They are similar to chasing tools except you use a hammer to push them along metal, and they cut and remove curls from the surface. To firmly hold the metal you should place it in pitch, use an engraver's block, or use a C-clamp attached to a jeweler's bench. Hold the chisel at an angle to the metal and strike its end as you push it along the surface.

The chisel pictured above was made for creating multiple lines. By turning it in different directions, a pattern can be created on the metal.

Etching
Etching metal is a great way to very quickly create detailed texture. To etch metal you apply a resist (something that resists the effects of acid) to its surface, and then put it into an acid bath. The exposed metal areas without the resist are "bitten" by the acid. (For a detailed explanation of the etching process, see page 114 of the Champlevé Enameling section.) When etching metal to create basse

taille textures, I use the same ferric based acids as for champlevé. Because basse taille etching doesn't have to be as deep as champlevé etching, the metal takes a shorter time in the acid bath. This allows you to use many more types of resists, such as permanent marker, fingernail polish, crayon, paint pen, or rub-on graphic design transfer materials.

Rolling Mill Textures

Rolling sheet metal with woven mesh

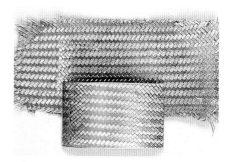

The texture embossed into the metal

A rolling mill is a large piece of equipment with two heavy steel rollers that can adjust to accommodate different gauges of metal. You can transfer textures onto metal by adjusting the pressure of the rollers, and rolling "sandwiches" of metal and embossing materials through the mill. You can use fabrics, plastics, and even paper cutouts to emboss metal, but for basse taille enameling, deeper impressions will show up best beneath the glass. Textures

Right: Joan MacKarell *Ram's Head Necklace*, 2003. 1⅗ x 1⅕ in. (4 x 3 cm). 18-karat gold, gold beads, calcite, enamel; engraved. Photo by James Austen.

Far right: Tom Reardon and **Kathleen Doyle** *Brooch*, 2003. 2 x 2 in. (5 x 5 cm). 18-karat gold, white sapphires, enamel; cast. Photo by Ralph Gabriner.

will be more visible through the glass if the embossing material is harder than the receiving metal. (Brass and nickel are harder than copper and silver.) You should also anneal the receiving metal so it embosses with ease. To protect the steel rollers from damage, keep hard metals such as nickel, titanium, or steel from coming in contact with the rollers by sandwiching them in brass or copper.

Engraving

Engravings are made with hardened sharp steel tools called gravers. You use the sharp blade of the tool to slice or curl away metal. The tips on gravers come in different shapes, such as triangular, half-round, or squared. Engraving well takes lots of practice. A sliding technique is used to create the smoothest and cleanest lines. To accomplish this, you press the graver into the metal at the start of a line, and push it along the line until the metal curls up and away from the surface. Skilled engravers can create beautiful basse taille surfaces on metal, and they also can cut away large metal sections with even walls to form champlevé recesses. A simple way you can use a graver is to make a zigzag line with the tool, rocking it from side to side as you move it along a line.

B

Joan MacKarell's engraving is demonstrated on this piece. In photo B, after the design was scribed on sterling silver, flat gravers were used to cut away the background area, showing the silhouette of the ram's head. After deciding which were the highest planes, different levels were cut to show the relief and model the piece. Once the modeling was complete, textures and details were added, such as the ram's horn and eye. The sterling silver was depletion-gilded before enameling.

Casting

Casting is a good way to texture metal. It's also an efficient method, because you can make a rubber mold of the original design and cast multiple pieces for different purposes. The brooch shown in the far right photo was cast from a rubber mold.

FAST FACTS & HELPFUL HINTS

— Always use clean, transparent enamel and test the colors on a sample to be sure the textures are clearly visible through the glass. You'll have the clearest and brightest colors if you sift out the fine grains and only use 80-mesh enamel. If you're creating a basse taille enamel on silver or gold, I suggest washing the 80-mesh enamel.

— Historically, basse taille surfaces were smooth and level. The color variations were simply a result of more enamel collecting in the recessed areas and creating darker values in the color. If you want to smooth the surface of a piece, use diamond sanding sticks or an alundum stone before the final firing. If you sand the surface with fine-grit carborundum papers or diamond papers, you'll be less likely to find pits in a re-fired surface.

— Transparent colors on basse taille have more clarity if you don't use organic binders to hold the enamel in place. If you're working on a dimensional surface and must use a binder, be sure to dilute it with water.

Fine Silver Foil & Fine Gold Foil

Sarah Perkins *Stitched Ceremonial Vessel,* 2002. 5 x 9 x 11 in. (12.7 x 22.9 x 27.9 cm). Copper, silver, 24-karat gold leaf, onyx, enamel. Photo by Tom Davis.

Gold AND SILVER FOILS enhance the brilliance of transparent enamel, allowing the rich color of the precious metals to show through the glass. Foils for enameling are very thin, fragile sheets of pure metal that can be purchased in booklets and easily cut and applied to an enameled surface. To prevent tearing or crumpling, handle single sheets of foil between pieces of tracing paper. Depending on the look you want to achieve, you can emboss the foil with texture, cut it with sharp scissors or a razor blade, punch it into shapes, or apply it in a mosaic pattern of small pallions. How you cut and apply foil depends on the size and design of its shape in your piece.

Here are important guidelines for using foil you should follow regardless of application method.

• Apply foil to a clean, even, and well-fired enamel surface.

• Use water to hold the foil in place. If you must use a holding agent, do so sparingly. If you work on vertical or sloped surfaces or use tiny pallions, you may need to use a little binder, but dilute it with water and use only when necessary. Gas is created when a holding agent burns out during firing. In later firings, the trapped gas can make its way to the surface.

• Always fire foil down to a base coat of enamel before you cover it with another layer of glass. If you apply foil to a base coat and don't fire it down before covering it with enamel, air will become trapped beneath the surface of the foil. During later firings, the gas and air will likely rise to the surface like a volcano, peeling back all the glass that is above it. In most cases, foil fired at approximately 1450° F (788° C) attaches to a base coat of enamel in about two minutes (or when the base coat fuses).

APPLYING GOLD & SILVER FOILS IN PALLIONS

This technique is useful when you need to apply foil inside intricate shapes. You can cut the pallions in any size, and though it can be a tedious procedure, it's a good way to fit foil into tiny areas. You can also use foil pallions with great control to create mosaic-like patterns for a textural effect.

Cutting & Applying Foil Pallions

1. Hold the foil inside a folded sheet of tracing paper. As shown in the photo, use scissors to cut tiny strips similar to fringe in the folded end of the paper-covered foil.

2. Over a clean piece of paper, use the scissors to cut across the fringed end, allowing the small pieces to fall onto the paper.

3. Wet a small paintbrush with water or water mixed with a drop or two of binder and barely moisten the area where you want to apply the foil.

4. Using the same paintbrush, pick up one foil piece and transfer it to the moist enameled surface. (Be careful not to pick up the tiny pieces of tracing paper that are mixed in with the foil.)

This sample (above) shows foil shapes that were once commercially available from enamel supply companies fired on a base coat of opaque blue enamel. Collection of the W. W. Carpenter Enamel Foundation. Photo by Keith Wright.

In the French text *The Painters' Enamel* (1866) Claudius Popelin explains that the use of gold, silver, and platinum foil gives a "vivacity and a brilliance of the most splendid effect." His method for cutting foils is explained in his own words as follows: "Place on a small board of pear tree wood, well trimmed and well polished, the sheet of metal which is to supply your foil; on the sheet of gold, of platinum, or of silver, place the vegetable paper where the outlines of the parts which you want to cut out are traced, and by means of a very sharp blade you will cut them by following your tracing in the most accurate and cleanest manner possible. A scalpel solidly set in a handle is excellent for this purpose. A small instrument formed of a thin shaft terminating by a very sharp iron tip of an arrow and a small knife for cutting leather sharpened on both sides also are very suitable. You will find these small tools at the vendors of surgery instruments."

Popelin goes on to explain another method of stamping small shapes explaining that, "it happens that one often has to employ small ornaments, the shape of which does not change, such as pearls, small crosses, fleur-de-lis, stars. Their small size would make it difficult to cut them out; therefore, they are executed with cutting tools, kind of steel

punches which one has engraved according to one's good pleasure, and which one keeps to use on occasion. For this, it is necessary to put the foil on a lead plate, cover it with a vegetable paper, and obtain a really clean cut by striking a firm and sure blow with a hammer or mallet to the head of the cutting tool."

Harold B. Helwig often incorporates foils with painted enamels. In the piece above he painted underglaze pigments directly on top of fired foil. He sifted transparent enamel over the piece and fired it until glossy.

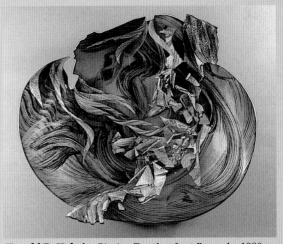

Harold B. Helwig *Piecing Together Lost Remarks*, 1990. 14 in. diameter (35.6 cm). Copper foil, underglaze painting, enamel. Photo by artist.

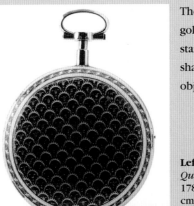

The peacock feathers on this enameled gold watch (left) are an example of a stamped foil technique. This pallion shape was frequently used on enameled objects in the late 18th century.

Left: France, *Enameled Gold Watch, Quarter Repeating à Toc,* circa 1786-1789. 1¹⁵⁄₁₆ in. diameter (4.9 cm). Enamel, gold. Bequest of Charles Phelps and Anna Sinton Taft, Taft Museum of Art, Cincinnati, Ohio.

Top: Kathleen Browne *Masked*, 2002. 3½ x 3½ x ½ in. (8.9 x 8.9 x 1.3 cm). Fine silver, sterling silver, 24-karat gold foil, enamel. Photo by artist.

Center: Jan Harrell *Icarus*, 1997. 4 x 2 x ¾ in. (10.2 x 5 x 1.9 cm). Copper, 24-karat gold foil, sterling silver, polymer clay, enamel.

Bottom: Tamar De-Vries Winter *Lovers' Cups*, 2000. 2⅕ x 2⅗ in. (5.6 x 7 cm). Sterling silver, gold foil, enamel. Private Collection. Photo by James Austen.

CUTTING & APPLYING LARGER FOIL SHAPES

1. Fold a sheet of tracing paper large enough to hold the foil sheet. Draw the foil shape you want to create on the tracing paper. Place the foil in the folded paper and hold it carefully, keeping the foil stretched tight.

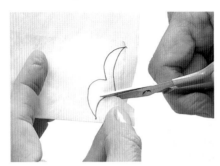

2. Following the drawing on the tracing paper, use small, sharp scissors to cut out the shape as shown, cutting the folded edge last. Let the foil shape fall onto a clean sheet of paper.

3. Use a dry paintbrush to manipulate and position the large foil shape on the surface of a dry enameled piece. Once the foil is in place, pick up a drop of water with a clean paintbrush and place it at the top edge of the foil as shown. The capillary action of the water will pull the foil to the surface of the enamel. Add another drop of water to an edge further down the foil, working it from the same end. Use a dry brush to gently smooth the foil as you go. Do not flood the enamel surface with water, and do

not add water drops from opposite ends or the foil will slide around the surface of the glass.

These samples by Harold B. Helwig illustrate the effects of foil pieces applied in layers. They also demonstrate how a gold-bearing transparent red enamel changes when fired directly on silver or over a clear coat of enamel on silver.

A

Photo A: opaque cobalt blue enamel was fired on both sides of this piece.

B

Photo B: 16 pieces of silver foil, each ½ x ½ inch (1.3 x 1.3 cm), were fired on the cobalt enamel surface.

C

Photo C: all 16 pieces of foil were coated with a transparent gold-bearing red enamel and fired. Nine additional ½ x ½-inch (1.3 x 1.3 cm) silver foil pieces were fired on the surface.

D

Photo D: the center foil square was covered with clear enamel, and the piece was fired. The piece was coated with a transparent gold-bearing red enamel and fired.

HOT TIPS

— You can use gold and silver foils together on the same piece, and you can slightly overlap foil pieces if desired. If different metal foils overlap, new alloys and a slightly different color will be created. You can apply additional enamel color over the foil after firing it down.

— You can paint underglaze pigment directly on the foil surface to create details before sifting transparent color over the piece and firing it.

EMBOSSING FOIL

Emboss gold and silver foils to produce a textured surface for basse taille enameling. Etched or roller printed metal works well as an embossing material. Even fabric or paper cutouts will successfully emboss thin foil.

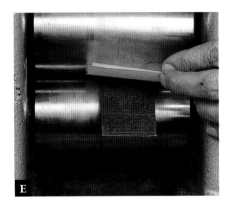

E

F

To imprint texture on the foil, you can run the foil and a textured material through a rolling mill (see photo E). If the foil is very thin, as most commercial sheets are, place it between sheets of tracing paper to protect it from tearing. Position the tracing-paper-protected foil next to the metal embossing plate, and then place a piece of leather or a paper napkin on top of the layers. Gently roll the whole "sandwich" through the mill so you don't tear the foil. Photo F shows the embossed foil (left) and the metal from which the texture was transferred (right).

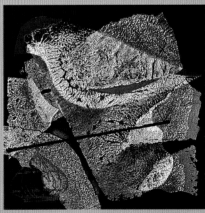

Top: Yoshiko Yamamoto *Brooch: After Gustav Klimt I,* 2002. 3 x 2⅛ x ¼ in. (7.6 x 5.1 x .6 cm). Fine silver, sterling silver, 22-karat gold, gold foil, pearl, enamel; fabricated. Photo by Dean Powell.

Center: Marilyn Druin *Shelter,* 2000. 2¼ in. (5.7 cm). Fine silver, sterling silver, 24-karat gold, 22-karat yellow gold, 18-karat yellow gold, enamel; cloisonné, guilloché. Photo by Mel Druin.

Bottom: Jenny Gore *The Heart Quivers,* 1999. 4 x 4 in. (10.2 x 10.2 cm). Gold foil, silver foil, enamel; reticulated, sifted, wet inlay. Photo by Trevor Fox.

An easy way to emboss foil is to hold it between sheets of tracing paper, place the paper against a textured metal surface, and press or gently rub the foil with a polished wood stick or even a pencil eraser until the texture from the metal imprints the foil. I encourage you to experiment to find other ways to imprint texture on foil. I've seen enamellists use a rolling pin, glass jar, and even their fingers to imprint foil.

Marilyn Druin *Goblet: Wedding Cup*, 1999. $5\frac{1}{2}$ x 2 in. (14 x 5 cm). Fine silver, 24-karat gold, 24-karat gold foil, enamel; embossed, cloisonné, basse taille. Photo by Bob Barrett.

FIRING FOILS

Once you apply foil to an enameled surface, allow the piece to dry completely. Fire it under normal conditions—about 1450° to 1500° F (788° to 815° C)—just as you would an enamel piece without foil. Remember, you're simply re-firing the coat of enamel that is beneath the foil. When appropriate, I occasionally inlay a little wet enamel on a different part of the piece so I can watch it fuse and know I have heated the piece to the correct temperature. With practice, you'll learn to recognize that when fired properly the foil stretches smooth. If it wrinkles or if an air bubble appears, prick the foil with a needle and re-fire it. Due to its high expansion rate in relation to the enamel, silver foil is especially prone to crack and rip from the surface of the enamel if left uncovered too long. To counteract this, you should apply the first coat of clean, transparent enamel very soon after you fire the foil down.

GOLD LEAF & SILVER LEAF

Although you can use gold and silver leaf with enamel, you should not use it in place of foil. You can tell the difference between leaf and foil by their weights. Leaf is so very thin that you can barely hold it without it falling apart. To use leaf as part of the final firing, simply stone the enamel surface, put on a holding agent, and apply the leaf. Leaf will cling quickly to a wet surface, and you cannot easily reposition it once laid down. Fire the enamel until the base coat glosses and the leaf is adhered to the surface. It will separate into fine cracks on the surface, revealing the enamel color beneath it.

FAST FACTS & HELPFUL HINTS

— Don't worry if gold foil comes out of the kiln looking dark or tarnished. Simply coat it with transparent enamel, and it will appear gold once again!

— Don't worry if your sheets of gold or silver foil look discolored or tarnished after being stored for a long period of time. After being fired and covered with enamel, they will look brilliant.

— Carefully select the transparent enamel colors you want to apply over foil. Remember how different colors appear over different base metals! Some colors may look better with a base coat of clear enamel fired over the foil. Reds and warm colors are spectacular when fired over gold foil and blues and cool colors are enhanced by the cool bright surface of silver foil beneath them.

Painting Techniques

Right: Barbara McFadyen *Brooch,* 2000. 2 x 1½ x ⅜ in. (5 x 3.8 x .9 cm). Copper, gold, enamel. Photo by Ralph Gabriner.

Far right: Martha Banyas *Brooch,* 1987. 2¼ x 2¾ in. (5.7 x 6.9 cm). Copper, silver, enamel. Photo by artist.

IT IS MY OPINION that of all the enameling techniques, painting with enamel requires the most skill and control. In this case, I am referring to painting with vitreous enamel, blending the ground glass in layers using opaque and transparent colors, to create tonal and color variations. Historically, this technique has been referred to as Limoges because it was in the city of Limoges, France, that the technique flourished in the 15th and 16th centuries. Studying these historical enamels in museums is for me a breathtaking and humbling adventure. The way the artists skillfully painted with glass is indeed impressive, and becomes even more so when you realize they were building and firing in wood-burning furnaces and making their own glass! Comparatively few contemporary enamelists paint with glass with the same detail and skill that was typical of true limoges enameling. If you practice blending and layering transparent glass, you'll be able to execute other wet inlay enameling techniques, such as cloisonné, champlevé and plique-à-jour, with greater control.

There are many simple ways for contemporary enamelists to work with painterly techniques because of the variety of new materials available commercially. Read this whole chapter before trying these techniques: it's essential that you clearly understand the differences between underglazes, regular enamel, (which in this case, refers to the vitreous glass we use for sifting) and overglaze painting enamels (which also include china paints called vitrifiable enamels, since they do not become vitrified glass until they fuse with heat). You can use these materials separately or in combination with each other, but you must first understand the application and firing procedures for each.

PAINTING WITH ENAMEL

Wet-inlaying regular powdered enamel was discussed on page 33 as a method of applying base coats or counter enamels. Painting with enamel is also a wet technique, done in a more painterly manner, blending and shading many thin layers of color to build value and dimension in the finished work. I'll begin by explaining the grisaille technique, which was really the first type of enamel painting, done so beautifully by the artists in Limoge during the 15th century. *Grisaille* comes from the French word *gris,* meaning gray, and grisalle enamels are primarily made up of gray tones. Limoges enamelists working in grisaille may have drawn their graphic inspiration from the portfolios of prints and illustrated books that were common at the time. They even shared their main drawing tool, the scribe, with printmakers.

Harold B. Helwig created this 6 x 6-inch (15.2 x 15.2 cm) sample shown above. He divided it into sections to show you the layering process for making a grisaille enamel. The first step is in the upper left hand corner. Following the sample counter clockwise, there are 11 individual coats of enamel added until the final finished section is complete. Look carefully as you move around the piece, and you'll notice that the gray values become lighter and lighter with each coat of white painting enamel.

Painting a Grisaille Enamel

1. Counter-enamel a metal piece. Sift and fire on a base coat of black enamel.

Long before the painting technique named after Limoges, France, was developed, the city's artisans had already developed a reputation for fine metalworking, goldsmithing, and champlevé enameling. In medieval champlevé, the areas to be enameled were scooped out of the metal and the depressions were filled with glass, as you can see in the Processional Cross on page 115. Though the same materials were used, the term "painting" is more descriptive of the technique that was developed in the 15th century and is now described as Limoges enameling. The outlines and areas of color were applied with a brush and spatula, and fused in layers in a series of firings. The technique was very painterly, and the artists who worked this way may not have originally worked as goldsmiths or metalworkers. It is thought that painting with enamel may have been discovered by accident when a monk experimenting in his workshop applied enamel to the surface of a piece without metal partitions, fired the piece long enough for the glass to fuse, and discovered that the colors did not run together. Limoges artists also discovered that applying a coat of enamel on the reverse side of an enameled piece prevented warping and cracking of the enamel on the front side. In the 15th century, Limoges artists first worked in the Grisaille technique, making pieces that were miniature paintings, usually with religious themes.

Workshop of Pierre Reymond
Standing Dish with Lot and His Daughters, mid-16th century. 3½ x 9⁷⁄₁₆ in. (8.9 x 24 cm). Enamel, copper. Bequest of Charles Phelps and Anna Sinton Taft, Taft Museum of Art, Cincinnati, Ohio.

Grisaille enamel plate, late 1700s. 8½ in. diameter (21.6 cm). Enamel. Courtesy of W.W. Carpenter Enamel Foundation. Photo by Keith Wright.

The enamel workshops in Limoges were family businesses and the art was passed down among generations. Leonard Limosin, Jean Penicaud, and Pierre Reymond were some of the most skilled masters at Limoges enameling. Most of the early enamellists looked at illuminated manuscripts, illustrated books, religious paintings, sculpture, and stained glass windows as source material for their designs. During the late 15th and early 16th centuries, the work of these enamellists gradually moved away from religious subject matter, and included portraits and narrative secular pictures. Leonard Limousin created some of the most beautifully rendered portraits ever executed in enamel. He was a painter and an engraver as well as an enamellist.

Léonard Limosin *Portrait of François de Clèves, duc de Nevers,* mid-16th century. 17¾ x 12 in. (45.1 x 30.5 cm) image; 29 x 23 in. (73.7 x 58.4 cm) framed. Enamel, copper, wood. Bequest of Charles Phelps and Anna Sinton Taft, Taft Museum of Art, Cincinnati, Ohio.

William Russell Birch
Lady Russell, 1783. 2½ x 1⁹⁄₁₀ in. (6.3 x 4.8 cm). Enamel. Courtesy of W.W. Carpenter Enamel Foundation. Photo by Keith Wright.

William Birch was born in England in 1755 and moved to America in 1794 where he continued to enamel. If he was not the first artist to enamel in America, he was certainly one of the first. Unlike the earlier work of Limousin which was painted with regular enamel, these portraits were painted over the enamel base coat using what we now call "painting enamels", or finely ground ceramic pigments mixed with glass that become vitrified when fired.

A Short Pink-Ground Enamel Bodkin Case, circa 1765. 4⅛ x 1 in. (10.5 x 1 cm). Enamel. Courtesy of W.W. Carpenter Enamel Foundation. Photo by Keith Wright.

2. Sift a thin layer of 150-mesh opaque antimony white enamel onto the base coat, and let it dry. (Different white enamels are made with different materials. Antimony white works best for the grisaille technique.) Scratch through the white enamel as needed to reveal the lines or areas you want to remain black. Blow or brush off excess.

3. Wet-milled white enamel is finely ground with water in a ball mill. If you prefer, you can mix this type of enamel with water and a holding agent or an oil binder. Simply let the binder soak into the glass. (If you stir the enamel mixture you'll create air bubbles.) This special type of enamel is sold commercially as "grisaille white," or you can also make your own. You'll need it for the next step.

4. Using a fine sable brush, paint the white grisaille enamel on top of the base coat in very thin layers to lighten the gray values. Fire the piece between layers. Subsequent applications and firings yield lighter values until pure white is reached in some areas (if desired). Firing the early painted layers very hot will soften the edges of the design and help the white fuse with the black to create gray values.

5. Paint the final details of your composition with a mixture of 325-mesh white enamel and a holding agent or oil binder. Fire this layer just long enough to fuse it.

Camaieu

Camaieu describes a painting technique very similar to grisaille enameling. Camaieu was more prevalent in the latter part of the 16th century. In this technique, painted layers of white enamel are built up on a transparent colored surface instead of a black one. Harold B. Helwig created these samples to show different stages of painting camaieu and impasto enamels.

Painting a Camaieu Enamel

• (photos A, B, and C) Apply antimony white wet-milled painting enamel to build up the white areas. The first application and firing may seem to almost disappear. Paint a second coat of white enamel onto the piece. This step will reveal the first white coat. Repeat the process to apply at least four additional coats of white painting enamel.

Impasto

• The impasto painting technique eliminates the need for so many firings. By painting the white enamel directly on the copper surface without a base coat beneath it, you can build up the layers more dimensionally in only one firing. (If you apply enamel this thickly and unevenly on top of a base coat, it would likely crack.) You can sift a transparent color over the white painting, and then apply successive layers of white enamel as desired.

Painting an Impasto Enamel

• (photos D and E) Soak antimony white enamel in a solution of one part holding agent and two parts water. Use a paintbrush to apply 325-mesh antimony white enamel to a clean copper surface. Sculpt the enamel in low relief, and let dry. Fire the enamel for 1½ to 2 minutes at 1470° F (799° C) until it is just attached. Do not over-fire the piece or bring it to a full gloss. Apply a black underglaze to the exposed copper surface. Cover the underglaze with a coat of transparent enamel, and fire the piece. Counter-enamel to finish.

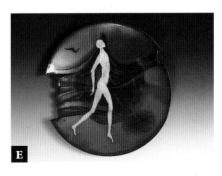

BLENDING COLORS WHEN PAINTING WITH ENAMEL

You can directly blend enamel colors into each other or use water and a small sable paintbrush to blend and fade them. You'll use the same painting technique described below to blend transparent colors for limoges, cloisonné, or champlevé pieces. (This sample piece has cloisonné wires outlining the design.)

A

• If the enamel is the correct consistency when applied, evenly blending two or more colors is not difficult. If the colors are too dry, they tend to clump up on the surface in lumps. If they are too wet, they will wash or spill across the surface. By tilting the paint tray as shown in photo A, you can pick up dryer or wetter enamel, depending on your needs.

B

• When blending together two enamel colors, I suggest blending the light color into the dark one (photo B). Repeatedly move the paintbrush up and down, pushing the grains of glass until you're satisfied with their appearance.

C

• If the blend doesn't look right, pick up more of the dark or the light color and continue dabbing it into the mixture (photo C). Grains of glass don't mix like paint, but they will mix visually as they lay beside each other.

D

• You can blend a single color from light to dark by adding water to thin it as shown in photo D.

• It isn't necessary to cover an entire piece with enamel before each firing. Work on specific areas and fire them to see if you're getting the value and color range you

desire. Proceed to add value or color, depending on the results.

E

• Thinly layer opaque, translucent, or transparent colors to create a painted enamel piece or cloisonné, as shown in photo E. Dry and fire the piece between layers until you're satisfied with the results.

• Don't blend wet enamel colors into areas that are already dry. This causes a separation known as *shorelining*, a flaw that looks similar to the line formed in the sand when the tides move in and out on the shore.

F

• Let the enamel dry well before firing.

Right: **Martha Banyas** *All We Are and All We Seem,* 1983. 26 x 40 x 5 in. (66 x 101.6 x 12.7 cm). Copper, brass, wood, enamel. Photo by artist.

Left: **April Higashi** *Community,* 2000. 3 x 2 x ¼ in. (7.6 x 5 x .6 cm). Sterling silver, 24-karat gold leaf, enamel. Photo by George Post.

When enamel colors dry they look chalky (see photo F), and you may think they aren't blended well, but remember how the colors looked wet and confidently fire the piece.

• When painting with standard enamel (regardless of mesh size) and blending colors, make sure the enamel is wet enough to easily move around the surface of the piece. If it seems to be clumping and not shading or blending easily, add a bit more water.

• It may be simpler to shade with 200-mesh or 325-mesh enamel, but to achieve the greatest transparency I suggest using the same painting techniques with 80-mesh enamel. When working over silver, gold, or textured metal, you should have no problems blending and shading an 80-mesh color.

• You'll have difficulty blending if a color is too saturated, such as a deep cobalt blue or a red. The grains of the deep color are always visible as specks in the lighter one. I reserve saturated enamel colors, apply them in their own firing, and don't often try to blend them with other colors. I do blend with satu-

rated color, but only by using water to shade and change their value. If I want to blend another color into a dark saturated color, I blend it during a separate firing.

WATERCOLOR PAINTING ENAMELS & ACRYLIC PAINTING ENAMELS

Woodrow Carpenter, owner of the Thompson Enamel Company, invented enamels that look and feel like watercolors, acrylic paints, and wax crayons. You can use these wonderful new materials in combination and with regular 80-mesh enamel. If you're familiar with other drawing and painting media but have never enameled before, you can quickly take advantage of these products—the only thing you need to learn is firing!

Note: If you're losing the reds and yellows when firing watercolor or acrylic enamels, you're probably firing too hot or too long.

Watercolor Enamels
Mi Sook Hur learned to enamel on the sample in photo G using watercolor enamels. Watercolor enamels are composed of finely ground enamel suspended in a

G

water-soluble base. To use watercolor enamels, wet a soft paintbrush with water and mix the color to a consistency you like. You can mix these enamel paints together on a watercolor tray just like ordinary artists' watercolors. Thinly apply the paint over a light or white base coat. For best results, fire many layers of color from light to dark to build up the values in the painting. If you apply them thinly and fire them correctly, the colors will be transparent, but you can also apply them more opaquely, depending on the effect you desire.

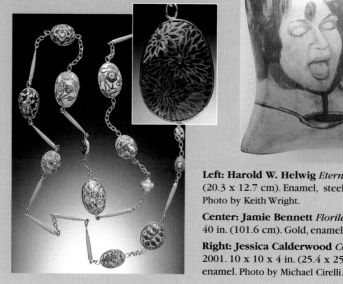

Left: Harold W. Helwig *Eternal*, 1976. 8 x 5 in. (20.3 x 12.7 cm). Enamel, steel; tinted grisaille. Photo by Keith Wright.

Center: Jamie Bennett *Florilegium Necklace*, 2003. 40 in. (101.6 cm). Gold, enamel. Photo by Dean Powell.

Right: Jessica Calderwood *Cough Syrup Dreams*, 2001. 10 x 10 x 4 in. (25.4 x 25.4 x 10.2 cm). Copper, enamel. Photo by Michael Cirelli.

Acrylic Enamels

Acrylic enamels are standard enamel powders dispersed in an acrylic polymer resin. You can mix these colors and apply them thinly (photo A), or use a brush or a palette knife to apply

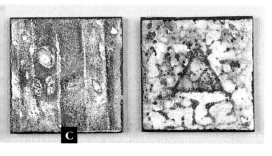

them more thickly. Acrylic enamels are very similar to watercolor enamels, and if one dries out in its tube, you can cut the tube apart and use the paint like watercolor. As with all enamels, you must let the painted piece completely dry before firing it. The samples in photos A, B, and C were made by Paul Hartley. On the two squares in photo C, he used acrylic enamels on silk gauze. He cut out the silk pieces and used regular acrylic medium to stick them to the enameled surface before firing.

UNDERGLAZES

Underglazes are primarily made up of metallic pigments, the colorants used in ceramic glazes and enamel, but they do not contain glass. When fired, they won't be shiny or glossy unless absorbed by the enamel layer beneath them or covered by a transparent coat of enamel (hence the name

underglaze, or "under glass"). The advantage of using these pigments is that you can very easily apply fine details, and they won't bleed or soften as much as regular enamel. Commercially available as pencils, ballpoint marking pens, chalks, and powdered pigments, underglazes offer great potential for working with familiar mark-making and painting techniques. Working with underglazes can be similar to using inks or thin washes. The samples in the first two columns on page 91 depict pieces created by artists with no prior enameling experience, but who are skilled in drawing, painting, and printmaking. The samples were all made on pre-enameled steel tiles and fired at 1450° to 1500° F (788° to 815° C) until the base coat turned glossy. In some cases, clear enamel was fired over the surface in a subsequent firing.

Catherine Walker created these samples using underglaze pencils.

The drawing was developed further with a second firing using both pencils and watercolors.

Watercolor enamels were applied over the drawing in a third firing.

Underglaze Pencils & Chalks

Ray Elmore created this enamel by drawing with underglaze pencils.

Scott Eagle drew with underglaze chalks to create this enamel.

You can buy underglaze pencils from ceramic supply companies. Potters frequently use them to create drawn imagery on glazed or bisque-fired clay. Although the chalks will give interesting effects if used on a wet glassy enameled surface, drawing with the chalks and pencils on a dry glassy surface is difficult, if not impossible. These drawing materials work best when you use them on a rough surface that has "tooth." To achieve this, you can stone, sandblast, or chemically etch the glass. Alternately, you could lightly spray the enameled surface with hair spray or a spray fixative. Both of these materials will later burn out when fired. After roughening the glass, you can use the pencils to create beautifully subtle shaded surfaces or linear areas.

Rebecca Laskin created the samples above using multiple techniques including sifting, stenciling, underglaze pencils, and pigments.

To determine the firing time and temperature of an enamel with underglaze pencils, keep your eye on its base coat. Fire until the base coat glosses, not the pencil lines. Depending on the amount of underglaze you apply, the piece may or may not need a cover coat of transparent enamel in order to gloss. Even if planning to apply a transparent cover coat, I like to fire the pencil work first to check the drawing. Sometimes a cover coat of clear enamel will soften the detail in the original drawing, but it can also give depth to the drawing once seen through a layer of glass.

Follow the same firing guidelines when you're working with underglaze chalks. If you fire the chalks down and they don't gloss or they look pitted on the surface, there wasn't enough glass to absorb the pigment. To remedy this, simply apply a coat of clear enamel or transparent color in the next firing.

Powdered Underglaze Pigments

A

Commercial underglaze pigments are sold as powders. You can mix them with water, a holding agent, or oil (whichever medium suits your application). I most often choose water, but it dries quickly. If you want to work for a longer period of time, try an oil-based medium. Use painterly techniques to apply them with a brush or quill pen. These pigments are the same as those found in underglaze pencils and chalks. In fact, you can use a razor blade to scrape dust off a piece of chalk to make your own powdered pigments (see photo A).

B

Prepare the powdered pigments in a paint tray that has small depressions. Use a tiny metal spatula or spoon to blend the pigment into the medium (photo B). Mash the powder with the liquid rather than stirring it. (Stirring creates air bubbles that might get trapped between layers of enamel.) I most often mix to the consistency of ink for painterly applications.

The sample pieces below, created by Barbara Minor, feature a pattern or texture using black underglaze with a tooled surface.

1. Mix black underglaze pigment with thinning oil to create a paste.

2. Apply the paste to the enameled base coat.

3. Use a flat color shaper (a rubber brush available from art supply stores) or a rubber eraser cut to the shape of your choice to create a design in the pigment paste.

4. Fire the piece until the design absorbs into the base coat and glosses. If the enamel doesn't completely gloss, re-fire it several times or apply a transparent cover coat, as shown in the sample above right. (You can apply the transparent cover coat over the design without first firing the pigment into the surface. Using a binder would smear an unfired surface, so if you work on dimensional surfaces, it's necessary to first fire down the pigment, as described above.)

In the samples above, Barbara Minor mixed a black underglaze with an oil-based medium and applied it to the recesses of an etched copper surface. Excess underglaze is wiped off the raised copper sections of the piece before it's left to gently dry on top of a kiln or under a heat lamp. When dry, transparent color is sifted on the surface and fired.

SAFETY MATTERS
• *Wear a dust mask when working with dry powders to keep from inhaling fine pigment dust.*

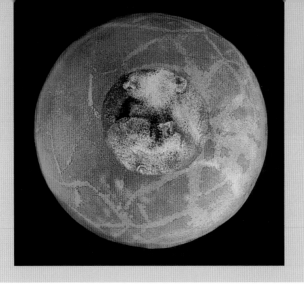

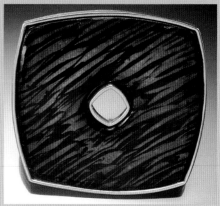

Right: Jessica Calderwood
Organ II, 2002. 20 x 8 x 8 in.
(50.8 x 20.3 x 20.3 cm).
Copper, enamel. Photo by
Michael Lundgren.

Far right: Barbara Minor
Brooch, 2002. 2½ x 2½ in.
(6.4 x 6.4 cm). Copper, 14-
karat gold, enamel. Photo
Ralph Gabriner.

Enamel Marking Crayons

Use enamel marking crayons on top of a glossy fired base coat of enamel, or on top of matte glass surfaces produced by sandblasting or stoning. A shiny surface can also be coated with hairspray or spray adhesive to give it more tooth and make the crayons easier to apply. This type of mark making is very similar to drawing with crayons on paper (photo C). The crayons are water-soluble. You can thin them with water and use them to make a wash. If the crayon is not fired and you're dissatisfied with your drawing, use a glass brush under running water to completely erase it. If you make thick crayon marks, they may seem pitted after firing because the crayons don't contain glass. A cover coat of clear enamel will smooth the pitted surface and give the piece more depth, since the colors will be seen through the clear glass. The samples to the left by Joan Mansfield show enamel marking crayons used on a pre-enameled steel tile with a sand-blasted surface.

OVERGLAZES, PAINTING ENAMELS & CHINA PAINTS

The use of overglazes and painting enamels differs from the painting techniques that originated in Limoges, France (though the term *limoges* is often used to refer to any painterly application of enamel). Those early artists used vitrified enamel. Overglazes, paint-ing enamels, and china paints are all finely ground powders that are not previously vitrified glass, but pure pigment with the addition of a finely ground flux. Whereas regu-lar enamel is fused before it is crushed and ground down to a particular mesh size, overglazes don't fuse and become glass until they are fired. Some artists grind regular enamel to a 325-mesh size so they can more easily apply it in a painterly way, and they may call this painting enamel. The differ-ence in terminology has to do with the manufacturing process of these products, but, in general, over-

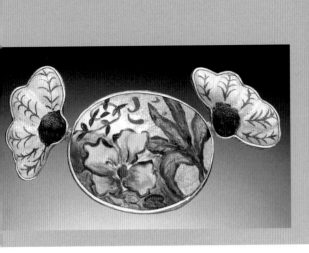

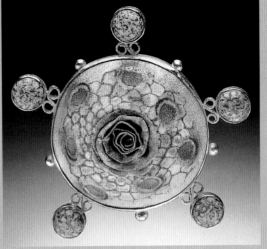

glazes are mixtures of finely ground pigments and low melting glass. Because the material is so fine (it feels almost like baking powder), it's possible to paint very fine lines and thin washes with great detail. Overglazes also fire at a lower temperature than regular enamel, approximately 1300° to 1450° F (704° to 788° C). You already may be familiar with china painting on porcelain. You can use these same china paints on fired enamel, and refer to them as overglazes.

If you fire regular enamel on top of a fired overglaze or china paint layer, the painting will likely diffuse or blur, burn out completely, or sink into the enamel surface beneath it, creating a gully. This happens because the pigment dissipates into the surrounding glass. You can sift a very thin layer of soft (low-firing) 100-mesh or 200-mesh clear enamel on your painting to protect it, but you still may lose some detail. If you sift larger grains of glass on top of an overglaze painting, it's more likely to disappear because large glass particles require more heat to fuse. To create a matte surface on top of an overglaze, try underfiring the transparent enamel coat to the sugar-fire or orange-peel stage. (Creating a matte surface through

chemical etching or sandpaper abrasion will likely remove the very thin painting enamel.) You can purchase overglazes in liquid or paste form, or as finely ground powders you can mix with water or oil mediums. Pre-mixed painting enamel is available in tubes. Similar to oil paints, they can be thinned with a little turpentine prior to use. Always mix and paint with overglazes in a well-ventilated area. The fumes are hazardous, and you should take care not to inhale the dry powder dust.

1. If you're using dry powdered pigments, put on a dust mask. Place a very small scoop of each color onto a glass surface or a ceramic tile (see photo).

2. Add a drop or two of oil binder (such as lavender or thinning oil) or water-based binder to the color. I prefer to use an oil binder because the resulting paint flows more easily and dries more slowly than it would if mixed with water.

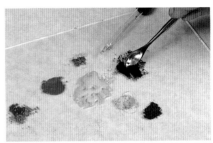

3. Use a palette knife to mash the powder into the binder. Pick up the color with the knife, and remix it against the glass or tile. Mix the paints well until they reach a consistency that is slightly thinner than toothpaste.

4. You can mix together two or more overglazes to create new colors. Prepare a palette of all the colors you wish to use. Keep some oil nearby to add to the colors as they dry out or to use for thinning the paints. If you need to stop working and want to save the colors, use plastic wrap to cover the glass or tile palette. Once oil-based colors completely dry out, clean the palette with turpentine and start over. If you use a water-based binder, only mix a small amount of color at a time because it will dry quickly.

5. It's important to use a good quality fine sable paintbrush to apply painting enamels. Work with smaller brushes to create fine

details and larger soft brushes for blending or stroking more color. You may want to use a clean soft brush without enamel to soften and blend colors. If you thin the paint to the consistency of ink, you also can use a crow quill or lettering pen to pick up color and apply fine lines.

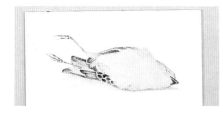

6. If you're blending painting enamels, it's best to work from light to dark as shown above in these examples by Beth Blake, a skilled painter with no enameling experience. Apply very thin coats of color and fire between the layers. It's more difficult to lighten an area that is already dark, than it is to darken a light area. Work rather quickly because as the mixture dries on the painting you cannot paint over it without lifting the color underneath.

7. Painting enamel should be completely dry before firing, or you can fire it soon after application if you "smoke" the piece to burn off the oil. To smoke a piece, quickly and repeatedly move it in and out of the furnace until smoke no longer comes off the surface (see photo A). You can then fire the piece at approximately 1300° to 1450° F (704° to 788° C), depending on the type of painting enamel used. Carefully watch the piece during the firing to see the surface change. You can put the piece back into the kiln if it's underfired. If you fire it too hot, the enamel will simply disappear or sink into the bottom layer of glass. If you've applied the painting enamel too thickly, the surface will look gritty even after it's fused. This quality may disappear if you re-fire the piece. As with all enamels, if you put the piece in a kiln that is too cool and bring it up to fusing temperature too slowly, cracks will appear in the painting enamel that are difficult to heal.

8. Continue firing the layers of painting enamel until you like the finished piece.

Metallic Lusters or Liquid Metals

Metallic lusters or liquid metals are fine particles of gold or palladium mixed with an organic medium, and you can use them as an overglaze. Enameling suppliers and ceramic supply houses sell these materials as liquids or in pre-filled writing pens. Apply metallic lusters thinly

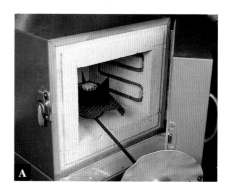

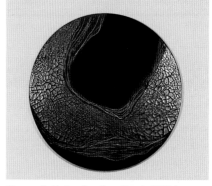

Kenneth Bates *Somber Depths Within*, 1955. 12 in. diameter (30.5 cm). Copper, gold luster, enamel. Photo by artist.

and with discretion in small areas. Liquid metals can be really difficult to fire if you apply them too thickly, so use a good quality sable brush, and paint with them evenly and sparingly. Dry them under a heat lamp or on top of a kiln before firing. Use the "smoking" method described in step 7 on this page to burn off the oils prior to firing if desired. Most lusters fire at about 1250° to 1300° F (677° to 704° C) for 1½ minutes or less. Do not overfire! After removing a piece from the kiln, check to see if the luster scrapes off, and re-fire it if necessary. Always apply lusters on a final firing. They will disappear if you place a higher firing enamel over them. Use lusters in a well-ventilated area as they may contain volatiles such as toluene, chloroform, and cyclohexanol which are very toxic.

HOT TIPS
TRANSFERRING A DRAWING TO METAL

If you're working from a detailed composition, you may want to draw your design on the metal or enameled surface before painting. Here are several ways to transfer an image.

— Draw it with water-base or permanent felt-tip markers. The ink from these pens will burn out during firing.

— Use a carbon paper technique. Depending on whether you're transferring a design onto black, white, or even unfired liquid enamel, you can choose different types of "carbon" paper. I often use red dressmaker's tracing paper that I buy at sewing supply and fabric stores. Because this is a waxy paper, it may interfere with underglaze pencils and chalks.

— Make your own "carbon" paper. Use a graphite pencil to heavily color the back side of the drawing you wish to transfer. Place the drawing right side up on the enameled surface, and retrace its lines. The graphite will leave a light line when fired.

FAST FACTS & HELPFUL HINTS

It's easiest to remember how to use the many materials I discuss in this section if you think of them in layers, all on top of a fired enamel base coat.

— Bottom Layer
Underglazes, enamel marking crayons, and ceramic pigments are the colorants you apply under a coat of transparent enamel. They do not contain glass and will not gloss unless they are absorbed into glass beneath or above them. Although you can use black underglazes directly on metal, in most cases you'll apply underglazes over a base coat of fired enamel. If you notice a pitted surface after firing these materials, fire a coat of clear enamel over them. This additional layer will usually absorb the pigment and create a smooth glossy surface.

— Middle Layer
Standard 80-mesh, 150-mesh, and 200-mesh enamels fire at about the same temperatures as acrylic and watercolor enamels. Determine the proper temperature and firing time by the particle size of the standard enamel. They do not require a top or a bottom enamel layer, but you can apply underglazes underneath

standard enamels, and overglazes on top of them. You can fire watercolor enamels and acrylic enamels next to each other because they fuse at about the same temperature as standard enamel. The red colors, however, seem to burn out more quickly, and you may need to fire them alone for a shorter time. Because watercolor and acrylic enamels have such a fine particle size they have less mass than standard sifted enamel and vitrify more quickly.

— Top Layer
Overglazes, painting enamels, or china paints are finely ground colors that you apply over the surface of fired enamel. Because of their small particle size, they fire at a lower temperature than standard sifting enamel. If fired too hot, overglazes will easily burn out or "sink" into a lower level of enamel. Do not attempt to stone or chemically etch the surface of an overglaze painting, or it may disappear. If you want your overglaze painting to have a matte finish, underfire a very thin sifting of low firing 100-mesh or 200-mesh clear enamel on the surface. Always use metallic lusters and liquid metals as a final firing.

Liquid Enamel Techniques

Julie Brooks Price
Communion, 2002. 6 x 13½ x 10 in. (15.2 x 34.3 x 25.4. Sterling silver, copper, steel, decals, liquid enamel, acrylic enamel, found vinyl recording blank boxes; limoges, screenprinted, cast, fabricated. Photo by Linda Darty.

LIQUID ENAMELS MAKE it possible to experiment with a variety of techniques related to ceramics, painting, and printmaking. Supplied commercially, this enamel is sometimes known as slush, crackle (because it cracks when fired over low expansion enamels), or water-based liquid form enamel. As described on page 35, the metal can be dipped, or the liquid can be painted or poured onto the piece. It can also be spattered, printed, trailed, pushed around, scratched through and manipulated in many ways. It's especially interesting to use liquid enamels on surfaces that have textures or folds so the liquids can pool in the low areas, creating tonal variations in the color. For all of the techniques and photographs in this book we have used lead-free water-based liquid form enamel in white, translucent white, transparent clear, red and black. Liquids are available in an array of bright colors that can be intermixed to create new colors. Ceramic pigments, which were discussed in the previous chapter, can be added to the liquids for an even greater range of colors. Because lead-free colors are acid resistant, the pieces can be dipped in pickle to clean firescale from uncoated copper surfaces. The firescaled areas can be recoated with liquid or powdered clear enamel so that after firing, they take on a reddish tone. There are many ways to experiment with liquid enamels, and in this chapter I've shared only a few that I find particularly appealing, or that my students have enjoyed.

POURING OR PAINTING LIQUID ENAMEL OVER FLAT OR TEXTURED METAL

I recommend you make a few test pieces to become comfortable applying and firing liquid enamels. You may want to use a dry or loaded paintbrush on flat or formed metal to create texture from the brush strokes. It's also fun to pour, dip, or paint liquid enamels on textured surfaces so the liquid collects in the recesses and creates tonal variations once fired. These effects can be especially interesting when you combine them with controlled over-firing. Practice firing white and clear liquid enamels at different temperatures to allow the copper oxides from the metal to come into solution with the glass and change its color. Where the enamel is thin, the color will be more green or blue, and where it is thick, the enamel will remain more true to the original color.

The samples on this page were all enameled on copper. Liquids in opaque and translucent white were poured or painted on the surfaces. To make the sample in photo A, liquid enamel was brushed on copper and the brush strokes were left visible.

Opaque white and translucent white enamels were used, and the pieces were fired at different temperatures.

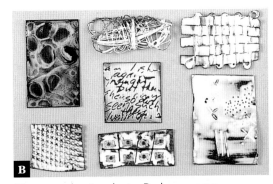

The samples in photo B show some of the diverse effects you can create by using liquid enamels over textured surfaces.

In 1761 the German technologist, Johann Heinrich Gottlob von Justy, suggested in the second volume of his *Collection of Chemical Writings* that ironware, glazed in a way similar to earthenware and pottery, would be much cheaper to produce than copper utensils. This eventually led to the production of "liquid enamel" or the addition of clay to suspend enamel in water. (The first patent for this process, however, was not recorded in England until 1839.) By the early part of the 19th century it was discovered that enamel could have utilitarian purposes, and so enameled pots and pans were manufactured. During the first quarter of the 20th century industrial enameling progressed to include the manufacture of stoves, refrigerators, kitchen sinks, bathtubs and home laundry appliances. The industrial enamellist efficiently coats thousands of identical pieces over and over again by dipping or spraying liquid enamel.

Top to bottom: blue enamel pan, circa 1920 (Courtesy of Woodrow Carpenter); gray mottled ware enamel pan, circa 1890 (Courtesy of Woodrow Carpenter); enamel spoons, circa 1920 (Courtesy of Linda Darty). Photo by Keith Wright.

The pots, pans, and utensils shown to the left are examples of some of the very early enamels that were made for utilitarian purposes. Mottled gray ware is one of the oldest known enamel finishes used on pots and pans. This pan dates from about 1890. In the "Proceedings of the First Annual Porcelain Enamel Institute Forum" in 1937, Mr. E. C. Dexheimer from the National Enameling and Stamping Company wrote comments about the process for making mottled gray ware: "The average enamel shop superintendent, whose entire experience has been based on stoves, tabletops, refrigerators, or white enameled kitchenware has no conception of the actual problems involved in producing this mottled gray ware. Owing to the fact that he sees this merchandise for sale in the ten-cent stores, we cannot blame him for assuming that it must be easy to make, in order to be able to sell it at that price. As a matter of fact, it takes more careful control to produce perfect gray ware than it does to produce any other enamel finish." He is of course referring to finishes on utilitarian pieces, typical of that period. The blue pan and the spoons pictured were probably made in the 1920s.

In 1973 Fred Ball wrote a book, *Experimental Techniques in Enameling*, to "enhance enameling with exploration evolving from traditional processes." Though he also worked with dry powder, it was his work with liquid enamels that opened the door for the type of experimentation still being practiced today. Some of his unique methods include spraying liquid enamel on cobwebs and firing the coated cobwebs onto metal; and coating found organic materials with liquid, applying them to metal, and then burning them out of the glass in the kiln. He threw enamel-coated string onto pieces, dropped liquid enamels from high off the ground to create splashed designs, and worked freely and with abandon, adding wax, sand, salt, cigarette ash, and anything else he thought might make an interesting surface on the glass. Fred Ball is best known for his experimental work but also for the large wall murals he made for public buildings in California. His investigations, though unconventional, were grounded in a thorough knowledge of traditional techniques. Fred Ball's work has had a great influence on the way that contemporary artists continue to explore interesting and unusual ways of working with enamel.

Fred Uhl Ball *Interior of Park Hills Place,* 1975. 40 x 32 in. (101.6 x 81.3 cm). Copper, enamel. Collection of the Crocker Art Museum, Sacramento, California. Gift of the Estate of Kathryn Uhl Ball. Photo by Kurt Edward Fishback Photography.

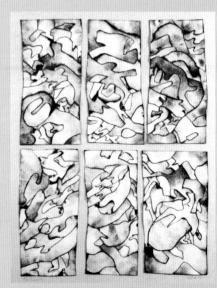

Fred Uhl Ball *Six Continents/Six Contentments,* 1976–1977. 36 x 27 in. (91.4 x 68.6 cm). Copper, enamel. Private collection. Photo by Kurt Edward Fishback Photography.

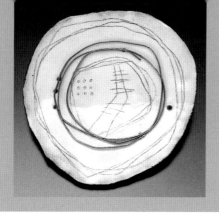

Right: Jan Smith *Which Way*, 2003. 1⅝ in. (4.1 cm). Copper, sterling silver, graphite pencil, enamel. Photo by Doug Yaple.

Far right: Iain Biggs *Eight Lost Songs*, 2003. 12 x 24 in. (30.5 x 61 cm). Steel, enamel (left); oil painted (right); enamel on copper. Photo by Elizabeth Turrell.

SGRAFFITO OVER LIQUID ENAMEL

A

Because of liquid enamel's fine particle size, it's simple to create finely detailed designs using the sgraffito technique (scratching through liquid enamel). You can paint or pour the liquid over bare copper or over a fired coat of enamel. When the liquid has dried to the correct consistency (experiment to see what works best for you), begin drawing through it with a sharp tool (photo A). Use a dry paintbrush to brush off the excess enamel you scratch off. If the enamel is too dry or the coat is too thick, it will flake off in large chips when scratched. Always wear a dust mask when blowing or brushing off dried enamel, and frequently damp-wipe your work surfaces.

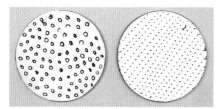

To create the textures shown in the samples above, Jessica Turrell scratched through dry liquid white enamel to expose bare copper.

The following samples were created by Michael Voors using the sgraffito technique in different ways.

B

Photo B: liquid white was poured on a bare copper surface. Once the enamel was dry, a needle tool was used to scratch the drawing through the enamel, revealing the copper beneath it. The piece was fired.

C

Photo C: black enamel was sifted and fired onto copper as a base coat. Liquid white enamel was poured over the black base coat and allowed to dry. A needle tool was used to scratch through the dried liquid white enamel to reveal the black undercoat. The piece was fired.

D

Photo D: white liquid enamel was poured over bare copper. The drawing was scratched through the dry enamel, and the piece was fired. Clear enamel was sifted over the surface and fired again.

Sgraffito Drawing in Layers

You can use multiple layers of sgraffito in combination with firescale and stoning techniques to create rich surfaces. The following samples by Elizabeth Turrell illustrate how this is done.

E

Photo E: this piece has a single layer of sgraffito. Opaque white liquid enamel was poured over the metal surface and dried. The drawing was scratched through the dried enamel, back to the bare copper, and the piece was fired once. Loose firescale that formed during firing was washed off the surface. Clear flux was sifted over the entire piece, and it was re-fired.

The following samples by Elizabeth Turrell (photos A, B, C, and D) show how a sgraffito drawing can change when created in layers that are then stoned or abraded to reveal lower layers of color.

Photo A: clear liquid enamel was poured over the copper surface and dried. A drawing was scratched through the enamel to the bare metal and fired. The dark lines that appear after the firing are firescale.

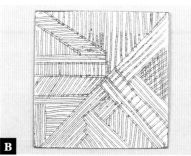

Photo B: white opaque liquid enamel was poured over the surface and dried. A drawing was scratched through the dried white enamel, and the piece was fired again.

Photo C: clear liquid enamel was poured over the surface, and the piece was dried. A drawing was scratched through the enamel, and the piece was re-fired. A thin second layer of white liquid enamel was poured over the surface, dried, and fired. The piece was stoned down to reveal areas of clear enamel underneath and then re-fired to create a gloss surface.

Photo D: white opaque liquid enamel was poured over the metal surface and dried. A drawing was scratched back to the bare copper, and the piece was fired. Yellow and red liquid enamels were painted on selected areas with a brush. The piece was dried, fired, and then lightly stoned with an alundum stone to reveal the drawing below.

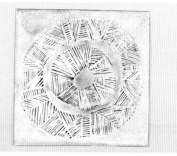

Photo E: to create this sample, Elizabeth Turrell fired black enamel on as a base coat. Black liquid enamel was painted on top of the base coat in circular strokes and dried. A drawing was scratched through the liquid black, revealing the black base coat, and the piece was fired. This created a surface with thick and thin areas of enamel, or high and low spots. A thin layer of white liquid enamel was evenly poured or painted on this surface, dried, and then fired. By stoning the uneven enamel surface with an alundum stone, the high black areas created in the original sgraffito drawing were revealed. The piece was re-fired to a glossy surface.

Photo F: this sample shows a similar technique. The difference is that clear liquid enamel was applied over the black sgraffito drawing before it was stoned instead of white liquid enamel. Also, the surface was not re-fired to a final gloss finish. The low areas were left shiny, and the high areas remained matte.

The following samples by Jessica Turrell shows other ways to use sgraffito to create beautiful fine details.

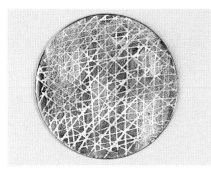

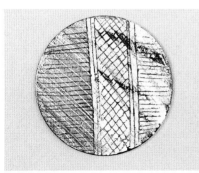

On this piece by Elizabeth Turrell, an image was drawn in the enamel through a stencil, and the piece was fired. Black liquid enamel was painted on the ends of the piece, and red liquid enamel was painted in the middle. The piece was dried. The image was re-drawn through the stencil, and the piece was fired. A thin layer of white liquid enamel was poured over the entire surface of the piece. It was dried, fired, and then stoned to reveal the underlying color and pattern. The piece was fired again.

PRINTING WITH LIQUID ENAMEL

As shown in this sample by Paul Hartley, you can make your own printing tools or use found objects to stamp liquid enamel or acrylic and painting enamels on top of a fired base coat. Liquids can even be printed directly on a copper surface.

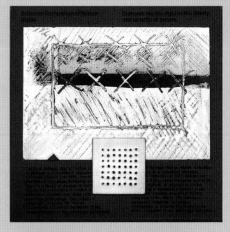

Top: Elizabeth Turrell *Universal Declaration of Human Rights Series,* 2001. 8 x 8 x ½ in. (20.3 x 20.3 x 1.3 cm). Steel, enamel; sgraffito, pierced. Photo by artist.

Bottom: Linda Darty *Garden Brooches: Winter,* 2002. 3½ x ½ in. (8.9 x 1.3 cm). Copper, enamel, silver. Photo by Robert Diamante.

Left: Linda Darty *Garden Cups: Winter*, 2004. 5 x 3 x 3 in. (12.7 x 7.6 x 7.6 cm) each. Copper, liquid enamel. Photo by artist.

Far left: Matthew Owen *Façade*, 2003. 1 x 1 in. (2.5 x 2.5 cm). Fine silver, copper, liquid enamel. Photo by Linda Darty.

PAINTING WITH LIQUID ENAMEL

A

B

You can use a variety of tools to apply liquid enamel in select areas. In addition to paintbrushes, try squeeze bottles (photo A), syringes, string dipped in liquid, your fingers, or anything else you can think of to spread the liquid. The left sample in photo B shows the fired piece. The sample on the right has a layer of flux fired over the entire surface.

CREATING TEXTURES WITH FIRESCALE, IRON FILINGS, SAND, LECITHIN, OR BORAX

In some cases, you may want your finished piece to have a gritty or grainy surface. You can accomplish this by adding materials to the liquid enamel before firing it.

C

Granulated borax, dust, cigarette ash, sand, or firescale residue are all materials you can thinly sprinkle onto unfired enamel surfaces to create texture. Experiment on small samples to discover what works best for you. Try many different liquid enamel colors as a base coat. Be sure to counter enamel the metal so the enamel on the front will be less likely to crack off. If you're using liquid enamel, apply it to the metal, and then sprinkle the texture material onto the surface while it's still moist (photo C). Let the liquid completely dry before firing. (If you're using sifted enamel, simply sprinkle the texture material on the surface before firing.) Borax

dust and ash will create pits and pinholes. Sand remains in the glass, giving it a rough tactile surface. The appearance of firescale changes little, remaining matte even after firing unless you coat it with a transparent color.

The samples in photo D by Gina Cox and Jennifer Hatlestad show a variety of materials fired into liquid enamel, including firescale, cigarette ash, sand, and gold foil.

ORGANIC OVERLAYS

You can coat a dry organic material with liquid enamel, let it dry, and then fire it on copper to create a skeletal shadow of the original form on the metal. Although this technique produces interesting results, it can be a bit messy and difficult to control. Dried leaves, cheesecloth, and cotton string or thread work well, but don't try to burn out materials that contain moisture or are too thick—they could explode!

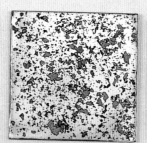

D

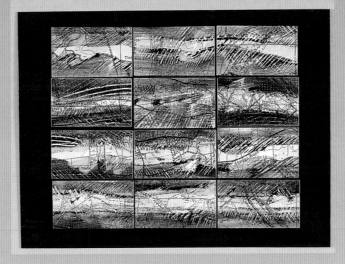

Right: Elizabeth Turrell *Memories & Maps: Landscapes of Childhood,* 2002. 15¾ x 12 in. (40 x 30.5 cm). Steel, enamel; sgraffito. Photo by artist.

Far right: Kate Cathey *Enameled Stickpins, Series One,* 2004. 5 x 2 x ½ in. (12.7 x 5 x 1.3 cm) each. Copper, enamel; fold formed. Photo by Robert Diamante.

Burning Out Overlays

1. Completely coat the dried material with liquid enamel, and let it dry thoroughly.

2. Place the overlay on a piece of clean copper and fire. The enamel will flame up, so keep the kiln door slightly ajar while the material burns out, and do not remove the piece until the smoke clears. (You should also be using a well-ventilated kiln!)

This is the finished sample by Dindy Reich with a coat of clear enamel fired on top of the burned-out overlay.

MARBLING, FEATHERING & SWIRLING LIQUID ENAMEL

If you manipulate more than one color of liquid enamel while the colors are semi-wet, you can create an effect similar to marbleized paper.

1. Apply a single-color coat of liquid enamel to a copper surface. Use a syringe or dropper to drip parallel lines or patterns of a contrasting color of liquid enamel across the semi-wet coat. Keep the coats of enamel very thin to prevent cracking and crazing.

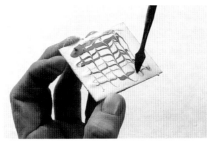

2. Using a feather or a pointed tool, draw through the lines of colored enamel to create a feathered pattern. If you tap, rotate, or move the copper as you work, you'll produce a more swirled or marbleized effect. Once you are satisfied with your design, let the piece completely dry, and then fire it.

The samples above by Dindy Reich illustrate the technique of feathering liquid enamel.

CRACKLE EFFECT WITH LIQUID ENAMEL

To achieve this effect, you first need to sift and fire a base coat of colored or transparent enamel on metal. Next, you apply a coat of liquid enamel on top of the base coat. Once you fire the piece, cracks will develop in its surface depending on the expansion and contraction rates of the base coat.

Photo E: clear enamel was fired on this sample by Matthew Owen as a base coat. White liquid enamel was poured over the base coat, dried, and then fired until the surface crackled.

Jessica Turrell *Sift Series Brooches*, 2002. 2-in. diameter (5 cm) each. Copper, silver, liquid enamel; sgraffito. Photo by Elizabeth Turrell.

Nancy Bonnema *Single Unit Spool*, 2002. 2½ x 1 in. (6.4 x 2.5 cm). Enamel, copper, sterling silver, binding wire, stainless steel cable, riso screening. Photo by Doug Yaple.

SIFTING DRY ENAMEL INTO LIQUID ENAMEL

Liquid enamel can function as a binder to hold dry enamel in place. An easy way to simultaneously fire a counter enamel and base coat is to paint a layer of liquid enamel on the back side of a piece, and then sift the counter enamel into it while it's still wet. Let the piece dry, turn it over, and sift a base coat of enamel on its front side. (Firing the counter enamel and the base coat at the same time makes the expansion of the metal and glass more equal and lessens the chance of warping.) Sifting dry enamel into liquid enamel also allows you to make a counter enamel that is thicker than liquid used by itself.

FAST FACTS & HELPFUL HINTS

— Keep mixed liquid enamels in lidded containers when not in use. If left uncovered, they will dry out and cake at the bottom of a container.

— When mixing liquid enamels, try to reserve some of the dry powder in case you need to thicken the liquid's consistency. If you make a mixture that is too thin and don't have any dry powder to add, leave the lid off the mixture, and it eventually will thicken as water evaporates.

— Wear a dust mask when you scrape or brush dried liquid enamel powder off a piece. This fine dust quickly coats work surfaces, and you could be breathing in the invisible, airborne particles.

— If you don't use a mixed liquid enamel for quite a while, it may become dried out and hard in the bottom of its container. To remedy this, put a little hot water on the surface of the dried enamel, and let it sit. Use a metal spoon to break up the enamel or use your hands to mash and squeeze out the lumps. You can even use a vertical hand mixer, such as the type used to make milkshakes, to blend the enamel. Eventually, you'll be able to remix the liquid enamel to the correct consistency.

— Use a well-ventilated kiln when burning out organic materials.

— Thoroughly prepare and paint the kiln floor with kiln wash so when enamel accidentally fuses there it can be easily removed. (For more information about using kiln wash, see page 18.) Better yet, you can place a broken kiln shelf on the floor to protect its surface when firing experimental techniques.

Cloisonné

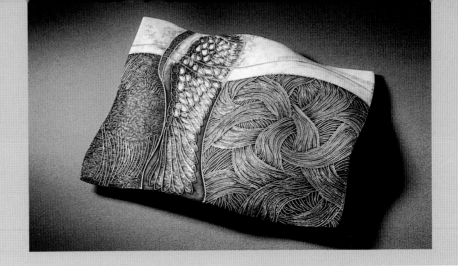

Fay Rooke *Gift of Time*, 1988.
28 x 25 x 3.2 cm. 24-karat gold and
fine silver cloisonné wire on raised,
pierced copper base with plique-à-jour
enamel. Photo by Terry Roberston.

CLOISONNÉ IS PROBABLY the most easily recognized enameling technique, and in fact, when I tell people that I'm an enamelist, one of the responses I often hear is "Oh yes, I know enameling, it's that technique that has all the little wires on the surface!" Cloisonné has a rich and long history, and because there are so many imported Chinese and Japanese items available for consumers to purchase, it's the technique most people identify as enamel work.

Cloisonné is a French word meaning "to be cut off." It describes a technique that utilizes thin ribbons of wire to separate areas of enamel color. Rather than soldering the wires to a metal base, as was the practice in Byzantine times, contemporary enamelists most often adhere the wires by fusing them into a previously fired coat of enamel. The design is arranged on the enameled surface with the flat wire standing on its thin edge, perpendicular to the piece and "glued" in place with a gum binder. Once the binder has dried, the piece is fired so that the wires barely sink into the base coat of enamel. Various colors of transparent, opaque, or opalescent enamels can then be inlayed around the wires and fired in layers until the glass is flush with the top of the wires. If you wish, the piece can be ground level and re-fired for a glossy surface, or sanded and hand polished to a matte finish.

CLOISONNÉ WIRE

Although you can make your own cloisonné wire, various gauges of wire in different heights and thicknesses are available commercially. Shorter wire (about 20 gauge) will result in a thinner finished coat of enamel. Taller wire (18 to 16 gauge) will require more layers of enamel in order to fill the cloisons, (but it also allows more opportunity for shading and layering color). Even wire of different thicknesses can be used in the same piece, and a thicker piece of wire can be hammered or milled through a roll mill so that when seen on the finished enamel, it will subtly change in quality, like a pencil line that goes from light to dark. Thin wire is usually about 34 to 36 gauge and thicker commercial wire is about 26 gauge. The wire can be formed with the fingers, pliers, tweezers, or around a jig to create a pattern or design that can include detailed cell-like areas, or simply open, linear designs. Fine gold or fine silver wire is most often used for cloisonné work, but copper or brass wire can also be used if you don't mind deal-

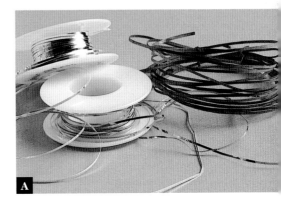

A

ing with the firescale that will form between firings. Copper wire is nice because it will oxidize to a black or dark color if you dip the piece in a liver of sulphur solution after it's finished.

SELECTING THE BASE METAL

Depending on how the piece will function, its size, and whether it will be formed or not, choose a base metal for jewelry size pieces that is anywhere from 24 to 18 gauge, and thicker gauges for larger pieces. Thinner gauges will need to be formed a little for structural strength to prevent warpage and cracking. Because it's more difficult to form the wires over a curved surface, a beginner might want to leave the piece flat, work on 18-gauge metal, or perhaps form a thinner gauge piece just minimally. If the design doesn't include very long lines, it's not difficult to form cloisonné wires over a domed surface.

HISTORICAL HIGHLIGHT

Cloisonné may have been the first enameling technique developed, long before others. In 1952, British archeologists excavated tombs in Kouklia, a small village in southwest Cypress that was an ancient city of wealth and importance in the late Byzantine Age. In one of the tombs, the archeologists found six gold rings decorated with cloisonné enamel that appears to have been fused.

Gold Ring with Cloisonné Enamel, 13th Century B.C. One of six rings discovered in a Mycenaean tomb at Kouklia, Cyprus in 1952. The earliest findings with cloisonné enamel in existence. Courtesy of Dr. Panicos Michaelides.

The very early use of cloisonné wires may have been inspired by the goldsmiths' familiarity with making enclosures for jewels out of soldered metal strips. When it was difficult to import various stones from far across the seas, metalworkers may have devised another method of including color by using glass. It is also quite possible that they thought the wires kept the glass from spreading, although with the glass available and the primitive firing methods, the early artisans had a difficult time even getting the enamel to gloss, much less spread very far! The wire cloisons probably did prevent the glass from cracking or chipping by dividing the piece into smaller enamel sections, adding strength to the metal, and reducing warpage.

Cloisonné began to be used quite differently by the Byzantine artists of the 10th and 11th century A.D. They used the wire to draw images with long flowing lines, creating figures and designs in their work that were narrative of biblical scenes or portraits of saints. The Byzantine metalworkers soldered the cloisonné wire to a very thin, fine gold sheet before inlaying the enamel. They created delicate and detailed work that marked the beginning of the use of enamel for stylized, figural art.

Enameling may have been introduced into China in the mid-14th or early-15th century from an Islamic source, and it seems enameling was introduced into Japan from China during the end of the 16th century. Because their culture was so separated from the West, most of the Japanese enamel artists in the mid 1800s had to come up with their own methods of working. In the Japanese cloisonné enamels of the 19th century, the wires were not soldered to the metal base. They were held in place temporarily by an adhesive, and then eventually by the fired enamel itself, similar to the way we do cloisonné today.

William Harper, an American enamellist who studied with both Ken Bates and John Paul Miller in the 1960s, developed a very personal way of working with cloisonné wire that did not involve tracing over a drawing or using the wire to simply create enclosed areas for enamel. He began to experiment with a more painterly approach to cloisonné enameling, using the line as a design element in the work and breaking free of the traditional constraints of descriptive or patterned design. Harper's cloisonné work continues to be expressive of ideas, emotions, and cultural comments, as he explores the artistic possibilities of the medium.

Above: Royal Gold Sceptre, 11th century B.C. 6½ in. (16.5 cm). Enamel, cloisonné. Discovered in a tomb at Kourion, Cyprus. By kind permission of the Nicosia Museum. Photo by Vassos Stylianou.

Right: *Medallion from an Icon Frame,* Byzantine, circa 1100. 3¼ in. diameter (8.3 cm). Enamel, gold; cloisonné. From the Djumati Monastery, Georgia (now Republic of Georgia); Made in Constantinople. The Metropolitan Museum of Art. Gift of J. Pierpont Morgan, 1917. Photo by The Metropolitan Museum of Art.

Japanese Cloisonné Enamel Vase, early 1900s. 8 x 6 x 6 in. (20.3 x 15.2 x 15.2 cm). Enamel; cloisonné. Collection of Linda Darty. Photo by Linda Darty.

William Harper *June Fragment,* 1991. 4⅝ x 2⅝ in. (11.7 x 6.6 cm). 14-karat gold, 24-karat gold, sterling silver, enamel, tourmaline, black pearl; cloisonné. Collection of the Racine Art Museum, Racine, Wisconsin. Gift of Janis and William Wetsman. Photo by the Racine Art Museum.

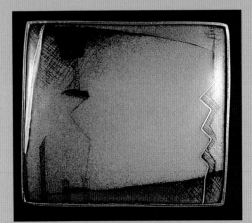

Rebekah Laskin Untitled, 1982. 1¾ x 1¾ in. (1.9 x 1.9 cm). Copper, enamel, sterling silver, fine silver; cloisonné. Photo by artist.

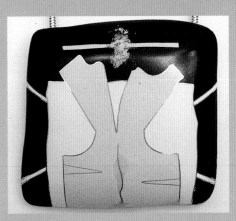

Jamie Bennett *Raglan Sleeve Pendant*, 1976. 2½ x 2½ x ¾ in. (6.4 x 6.4 x 1.9 cm). Enamel; cloisonné. Collection of June Schwartz.

"If you treat lines as barriers rather than transitions that describe an area of space, you are limiting yourself a great deal. A line is a visual element, not a technical one. Allow yourself the opportunity to think about the measure of a line, its width and length; the type of line, be it curved, straight, or angular; and finally its expressive quality, which gives life to the line. Once you admit the natural qualities of line in your work, you will find a freshness which is not limited by any process, but enhanced by your awareness."

Jamie Bennett

APPLYING THE BASE COAT OF ENAMEL

If you use a light-colored base coat, you'll be able to work with layering transparent colors more easily because it's simpler to work from light to dark. If you're layering with opaque enamel, the base coat color is not so important. If you're working on copper, I suggest using a light white or off-white base coat or perhaps clear enamel. If you work on fine silver, you can apply a clean, clear enamel as the base coat. Sift the fine enamel out of the clear enamel, and you may not need to wash the remaining 100-mesh or 80-mesh enamel. If you choose a dark base coat, you might enjoy firing foils onto different areas of the piece, so you'll have some light areas to contrast with the dark color.

FORMING CLOISONNÉ WIRE

You may find that cloisonné wire needs cleaning or has become crumpled and bent in storage. To clean and straighten the wire at the same time, you can pull it between a folded green scrub pad (photo A). Alternately, you can straighten wire

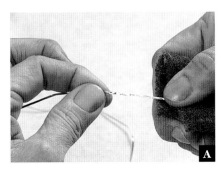

by firmly placing one end in a vise and stretching it with pliers (photo B). To make the wire easier to bend or to form over a three-dimensional surface, you may need to anneal it. Place the coiled wire on a clean firing screen or mica sheet and heat it in a kiln until the trivet or screen glows dull red (photo C). If you're annealing fine silver or gold wire, quench it in water. You must pickle copper or brass wire after annealing to remove firescale.

Forming Cloisonné Wire over a Drawing

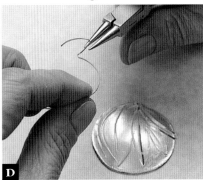

Depending on the design, you can form cloisonné wire with your fingers, pliers, or tweezers (photo D). If you intend to create closed wire

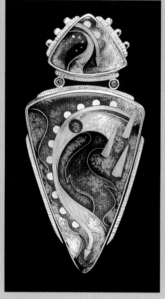

cells, make sure the wire ends butt up against each other.

A

If you're working from a very detailed drawing, it might be simpler to form the wire over double-stick tape. This method prevents the wires from slipping and moving as you follow the lines of the design. Better yet, try placing a piece of glass with polished edges over the drawing, and then putting the double-stick tape on top of the glass. This allows you to easily move your work off the drawing to check it and easily slide it back over the drawing to continue working on the wires (photo A).

Beading Cloisonné Wire

B

Cut tiny lengths of fine silver or gold cloisonné wire. Heat them to create metal "beads" that are

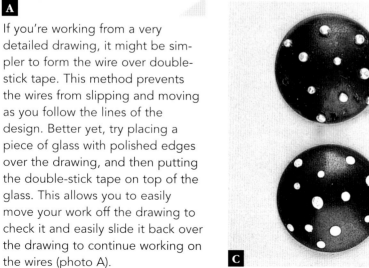

C

about the same height as the cloisonné walls as shown in photo B. (You might have to practice with a few different lengths until you make the right size "bead" to fit flush with the cloisonné wire.) When you inlay the enamel around the beads and stone it flush, the beads will appear as solid circles of metal (photo C, top sample). You also can place tiny metal beads on an enamel prior to its final firing, and the beads will remain above the surface of the glass (photo C, bottom sample).

Twisting Cloisonné Wires

D

Use twisted cloisonné wire to create dotted lines on an enamel surface. Secure one end of a cloisonné wire in a vise and use pliers to twist the opposite end (photo D). Fuse the twisted wire to the enamel surface (photo E), and then apply an

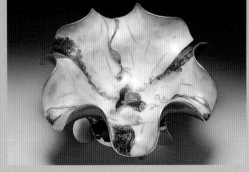

Above: Erica Druin, Marilyn Druin, and **Michael Good** *Emerging Legacy,* 2001–2002. 3¾ x 5¾ x 5¾ in. (9.5 x 14.6 x 14.6 cm). 24-karat gold, fine silver, enamel; cloisonné, basse taille. Photo by Bob Barrett.

Right: Marilyn Druin *Aquamarine,* 1998. 2½ in. (6.4 cm). Fine silver, 24-karat yellow gold, 18-karat yellow gold; cloisonné, basse-taille. Photo by Bob Barrett.

Far right: Frances Kite *Depths of Beauty Brooch,* 2002. 2⅜ x ⅝ x ¼ in. (6 x 1.6 x .6 cm). Fine silver, silver foil, gold foil, enamel, garnet, cultured pearl; embossed, cloisonné. Photo by artist.

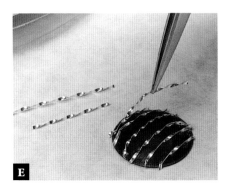

Making Straight Lines with Cloisonné Wires

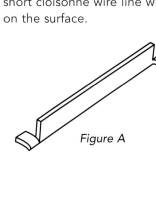

opaque color around the twisted wires. The low areas of the twisted wires will disappear beneath the enamel (photo F).

To prevent cloisonné wires from falling over in the kiln, each one must be bent so it will balance on its thin edge. One way of making straight lines is to double the length of the wire and extend it off the edge of the piece for balance (photo G). Once the enameling is complete, you can use nail clippers to cut off the excess wire (photo H).

Forming Short Wire Lines by Cutting a "Foot"

Make short lines by cutting a "foot" in the wire as shown in figure A. If you use an opaque enamel to cover the foot, only the short cloisonné wire line will show on the surface.

Figure A

Forming Wire with a Jig

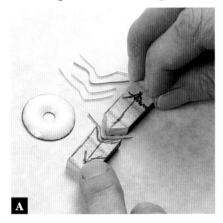

A

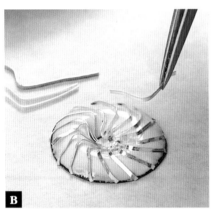

B

Corey S. Fong *Spiral Movement*, 2003. 1¾ x 5 in. (6.9 x 12.7 cm). Enamel, silver, copper, garnet, pearl; cloisonné, reticulated, forged.

You can make a tool that will allow you to easily create the same cloisonné shape repeatedly. Corey Fong simply cut this piece of wood and used it to press the zigzag cloisonné shapes in this piece (photos A and B). Place the correct length of cloisonné wire in the jig, and then squeeze the wood pieces shut. As shown in photos C and D, you can also produce a more stable jig that you can adapt to make many different wire shapes. Harlan Butt designed and constructed this example. The brass piece that bends the wire can be removed and replaced with other cut brass pieces.

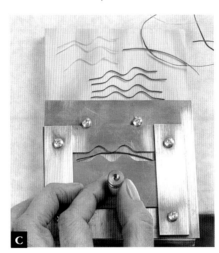

C

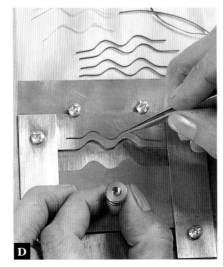

D

ADHERING THE CLOISONNÉ WIRE TO THE BASE-COATED METAL

E

F

If you're working on a flat or slightly domed structure, you can use a holding agent to adhere the cloisonné wires to the base coated piece. Dip the wires in the binder before transferring them to the piece as shown in photo E. Sometimes this method causes the wires to float or move around on the surface of the enamel, and it may be difficult to control a very intricate design. If so, simply paint a smaller amount of holding agent around the wires as you apply them to the piece, as if "gluing" them down (photo F).

To apply cloisonné wires to three-dimensional pieces with vertical surfaces, you'll need to use a different binder. Some binders are available from enamel supply companies for this purpose, and my favorite is made from lily root powder or lotus root powder. To mix it, sift the powder into a shallow container of water until the powder covers the surface of the water (photo G). The formula will thicken and form a strong binder into which you can dip cloisonné wires before placing them on the dimensional piece. Once the binder is dry the wires will be so tightly adhered to the enamel surface that they may be difficult to remove.

ENAMELING A CLOISONNÉ PIECE

The following samples by Harlan Butt illustrate one way of making a cloisonné piece. Because the artist wanted to apply the enamel color over silver foil, one large piece was applied over the entire base coat. A coat of clear enamel for silver had to be applied over the foil before applying the wires because cloisonné wires will not adhere to silver foil. If you wish to fire foil down in select areas instead of over the entire surface, do this in the first or second firing after the cloisonné wires are attached. (Remember that you must com-

pletely fire down foil before coating with enamel. Air bubbles beneath the foil can cause it to rise to the surface and break through the enamel. See page 84 for more information on firing foils.)

1. Cut a disk of copper, slightly dome the form, and clean it well. Apply and fire a base coat of hard clear enamel on the copper disk as shown. (If you plan to cover the entire piece with silver foil, you can use any high-firing enameling color as a base coat. If you use a soft-firing base coat, the cloisonné wires could sink through to the base metal, or the base coat might bleed through the enamel layers fired above it.) Coat the back side of the disk with a counter enamel and fire. You can apply more than one coat of counter enamel if you plan to fire many enamel layers on the front side of the piece.

2. Carefully place silver foil over the top of the disk. Wet the front of the disk with a little water so the foil is easier to position, or you can apply it dry as described on page 82 and paint water under one edge after the foil is in place. Wrap excess foil around the back of the disk or trim it off even with the edge. Let the disk dry, and then fire it to adhere the foil to

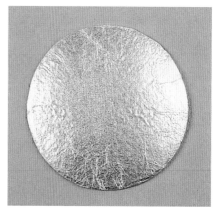

the enamel base coat. After cooling, sift a thin coat of clear enamel over the silver foil.

3. Bend cloisonné wires into the shape you desire and arrange them on the disk. Adhere the wires to the enamel by either dipping them in a holding agent or painting a holding agent around them as they are placed. Let the finished piece dry well.

4. Fire the wires down at the correct temperature for fusing the base coat. Watch the piece carefully and do not over-fire it! (I find it easiest to watch the trivet: when it turns a dull red color, check the piece.) Remove it from the kiln and use a palette knife or spatula to gently press the wires, checking that they're in place. Work quickly and carefully while the enamel is still hot enough to catch the wires.

You can fire all the wires onto the piece at once or you can fire some in place, and then add more in additional firings.

5. Using the wet inlay technique described on pages 33–34, fill the areas around the wires with enamel. Work slowly and carefully. If you're using transparent colors, thin coats of enamel will produce better clarity than heavy coats. Through sequential firings, build up the enamel in thin layers as necessary until flush with the top of the wires (see page 88 for a demonstration of this process). Inlay a little more color around the edge of each cell so that as the enamel

pulls toward the center it will be more level (see figure B). Sometimes I fire a piece as many as 20 times to achieve this result.

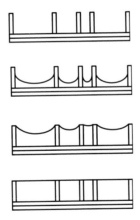

Figure B

6. Once the enamel is flush or nearly flush with the wires, grind the fired surface with an alundum stone or diamond sandpaper to create a smooth level surface. (For more information on grinding, see page 52.) Using only mild pressure, grind the piece under running water until the wires are fairly level. You may want to refill any low areas with enamel and re-fire

the piece, rather than continuing to grind it, which will take off more layers of color. (You could also use a lapidary rubber-drum sanding machine to level the wires and finish the surface, but be careful! It's easy to abrade too much color too quickly when using this method. Refer to page 53 for further information on using a rubber-drum sander.)

7. Once the cloisonné surface is completely level, clean it well with a glass brush. Re-fire the piece to heat-polish it to a glossy surface or hand-polish it for a matte finish as described on page 52. File and polish the edges of the piece.

Right: Harlan Butt *Horizons: Olympia #1*, 2004. 6 x 9 x 7 in. (15.2 x 22.9 x 17.8 cm). Silver, enamel. Photo by Rafael Molina.

Far right: Jessica Turrell *Cloisonné Box*, 1997. 3$\frac{1}{8}$ x 3$\frac{1}{8}$ in. (8 x 8 cm). 9-karat gold, sterling silver, enamel, fine silver wire; cloisonné. Photo by James Austen.

FAST FACTS & HOT TIPS

— If you have difficulty forming cloisonné wire, try annealing it to soften the metal before you begin. It may also be helpful to re-anneal the wires after forming them and before firing them onto the piece.

— Try forming cloisonné wires over double-stick tape. Following a complicated drawing is much easier if the wires are held in place as you work. Apply the double-stick tape directly to the tracing paper drawing, or to a small piece of glass placed over it.

— Position your worktable at or near eye level while you form wires so you don't strain your neck or back by bending over for too long. Stand up and stretch often!

— Before placing wires on the enameled surface, you may find it helpful to partially grind the base coat with an alundum stone or diamond sanding stick. This will give you a visual indicator that the wires are fired down. (When the ground area becomes shiny, you know the base coat has fused.) Just make sure to grind an area that is large enough to see. If you can't see the rough surface in the kiln, quickly look at it when you remove the piece from the furnace. If it isn't shiny, you'll know to put it back in

the kiln. I usually use the color of the trivet (dull red) as an indicator that the base coat has fused and the wires have attached.

— If you over-fire the piece, the cloisonné wires will sink too low. If fine silver sinks down to copper it creates a new alloy known as a eutectic, which melts at a lower temperature than either fine silver or copper. The eutectic will appear black and mottled. If you are afraid of over-firing a piece, take it out at the orange-peel stage each time, until the final firing.

— When wet inlaying a cloisonné piece apply your chosen colors early in the layering process. If you wait until the final few firings to make the piece look the way you want, you may find that you lose those layers during grinding.

— If you're working on fine silver, you should apply clear enamel wherever you plan to inlay a warm color. For example, if I were making a green to yellow blend, I'd apply the darkest greens, middle-value greens, and lightest greens, and put clear enamel in the places that will eventually be yellow. After firing this layer, I'd inlay the yellow over the clear enamel.

— If you're concerned that you may grind or sand through the colors as you level the enamel with the wires, you can apply well-washed clear enamel as a final layer. If you've stoned or sanded your piece after your final firing and you find low spots that need to be filled, you can inlay well-washed clear enamel or a color into those low areas and re-fire the piece.

— If the enamel reaches the top of the cloisonné wires and you realize that you'd like to change or add a little more color, you can! Simply use a small round diamond bit in a flexible shaft machine to remove the color you want to replace. Wear a dust mask and be sure to wet the enamel so the diamonds won't come off the bit. You can even use the wet diamond bit to blend colors by grinding and softening edges. Be careful not to damage the cloisonné wires during grinding.

— Early in the wet inlay process, if you decide you don't like a color, you can apply foil over it. Be careful, though. If you wait too late to add the foil, you might end up grinding it away when you level the surface.

Champlevé

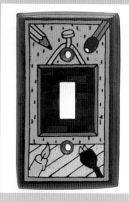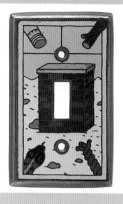

Pencil Brothers *Set of Switch Plates*, 1975. 4½ x 3 in. (11.4 x 7.6 cm). Copper, enamel; champlevé. Photo by Lynn H. Thompson.

CHAMPLEVÉ ENAMELING is one of my favorite ways to enamel because I like the quality of the lines that are easy to create using a variety of different techniques. The term *champlevé* comes from the two French words "champ," a field, and "levé," raised. In this technique raised fields, or areas of metal, are incorporated into the finished design and the enamel is inlayed into recessed compartments. During the Middle Ages, artisans chiseled and gouged out surfaces of metal and filled them with enamel, sometimes using cloissonné line work in the recessed areas as well. Contemporary artists use engraving, and also acid etching, embossing, electroforming, and fabrication techniques to create the recesses into which the enamel is inlaid. In this chapter, photographs and short descriptions of a variety of techniques are presented to help you differentiate between some of the many ways you can do champlevé enameling. Detailed instructions are included only for acid etching because it is simple to do, even for a beginner, and it has such a range of design possibilities. In the project section of this book there is also detailed information on how to make a brooch using what I call the "saw and solder" technique for champlevé, an efficient way to work on small pieces for someone with sawing skills.

Etching for Champlevé Enameling

METALS

Traditionally 14- or 16-gauge metal is used to etch pieces for enameling, especially large pieces, because the heavier metal will still have structural strength, even after it is etched. If you want to have plenty of room to inlay color, you should plan to etch about halfway through the metal, but this isn't always necessary. On large vessel forms, use 14- or 16-gauge metal, but on smaller jewelry pieces, especially if they are formed, 18- or 20-gauge metal works well, depending on the function of the piece.

ACIDS

Though nitric acid can also be used to etch silver and copper, I prefer ferric nitrate and ferric chloride because they are readily available, safer to use, bite the metal more cleanly, and can be used with a wider variety of resists. Nitric is more hazardous to work with, doesn't bite the metal as cleanly, and limits the resists that can be used: nitric acid is highly volatile, bubbling and lifting resists during the etching process. The disadvantage of the ferric acids is that they etch so slowly, but it is worth the extra time because the slow bite not only gives a cleaner etch with less ragged edges, it also makes it

possible to use more resists. With these ferric-based acids you can use many resists for shallow etches (such as crayon, permanent marker, glue, etc.) but my favorite resists for deep etching are an oil-based paint pen or marker and circuit board heat transfer paper (see page 118).

Ferric chloride, used straight from the bottle, is inexpensive and readily available from radio parts stores. It can also be mixed from crystals (13 ounces [340 g] of ferric chloride etching crystals to 16 ounces [.5 L] of water.) Ferric chloride will etch copper, but not silver. A mustard colored residue is left on the surface of copper when etching with ferric chloride and it should always be neutralized and scrubbed off the metal with ammonia or ammonia and water after etching.

Chariot Trappings, Provincial Roman, perhaps Thrace, 200–400. Largest plaque: 5⅝ x 3⅞ in. (14.3 x 9.7 cm). Copper alloy, enamel; champlevé. The Metropolitan Museum of Art. Purchase, Jeannette and Jonathan Rosen Gift and Fletcher Fund, 2000. Photo by The Metropolitan Museum of Art.

there is evidence of warriors' swords, shields, and horse trappings using the champlevé technique.

As champlevé enameling reached Europe, it became the predominant technique in Limoges, France, at the turn of the 12th to the 13th century. Lay establishments began to create commercial champleve enameled goods, near the monastic centers around Limoges. These ceremonial and religious objects were exported throughout Europe until the late 14th century. Workshops produced such items as incense burners, crucifixes, crosiers, reliquaries, portable church altars, caskets, and bookbindings. Deeply glowing opaque enamel colors were applied to richly gilded metal. The champlevé enameling workshops in Limoges, France, were destroyed by the troops of the Black Prince in 1371,

Celtic craftsmen used champlevé techniques long before the Roman expansion into Western Europe. How the Romans eventually gained access to this form of enameling has not been documented, but during their land conquests they may have come into contact with cultures that used enameled objects. A creative exchange may even have taken place as the Romans were inspired by the Celtic objects they saw, while the Celts may have been influenced by the Roman glassmaker's works. The Celts began enameling in the 1st century B.C. and by the end of the 5th century A.D.

Processional Cross, Limoges, France, circa 1225–1250. 13¹³⁄₁₆ x 8¹⁄₁₆ in. (35.1 x 20.5 cm). Enamel, copper; champlevé. Bequest of Charles Phelps and Anna Sinton Taft, Taft Museum of Art, Cincinnati, Ohio.

during the war between France and England. It was not until after the reign of Louis XI (1461–1483) that the revival of enameling in Limoges took place and the painting techniques that the city is so famous for developed.

Enameled buttons were not mass-produced for women's clothing until after 1870 and they continued to be popular through the first quarter of the 20th century. The pieces were usually die stamped using machinery so that multiples could be made, and the colors were all applied by hand. In addition to champlevé buttons, champlevé badges can still be found in the collections of organizations with political, social, and sporting interests. Changes in the 19th century gave manufacturers access to steady supplies of enamel, copper-alloy sheet, economic fuel for firing, and also the machinery and laborers common during the industrial revolution. Small family-run factories produced coins, buttons, jewelry, and badges, which used stamping, embossing, or casting to make the raised and recessed planes in the designs.

John Paul Miller was still a student in high school when he enrolled at the Cleveland Museum of Art school and studied with Kenneth Bates. He eventually began making rings and brooches in his parents' basement, but he also worked on watercolors and thought that was the field he would pursue. After World War II, Miller began teaching at the Cleveland School of Art and making jewelry in gold. The form in this piece was made from pure gold to get the best color, and 18-karat gold pieces were fused to the pure gold form before enameling. The fusing is a technique known as *granulation.* The fused pieces create raised planes on the surface, and can be described as a type of champlevé.

Enameled Buttons, late 19th–early 20th century. Various dimensions. Courtesy of W. W. Carpenter Enamel Foundation. Photo by Keith Wright.

John Paul Miller *Snail.* 2¼ x 1⅛ in. (5.7 x 2.8 cm). 24-karat gold, 18-karat gold, enamel. Photo by artist.

Ferric nitrate is the acid I use for etching silver. It etches very cleanly and also works well with an oil based paint pen and heat transfer paper. It will also etch copper, but ferric chloride is much less expensive and seems to last longer. 500 grams of ferric nitrate crystals are mixed with 1 pint (.5 L) of hot water until they dissolve (or 3136 grams to 1 gallon [3.8 L] of water if you use a spray etching machine). Ferric nitrate is basically nitric acid, which contains iron and it is neutralized with baking soda.

Kristin Anderson *Feather Pins*, 1998. 3 in. (7.6 cm) each. Sterling silver, enamel. Photo by Thomas Baird.

SETUP FOR FERRIC CHLORIDE & FERRIC NITRATE

During etching, both of the ferric-based acids leave a layer of sediment on the exposed copper. As the acid eats the metal, the residue sits on the metal surface, creating its own resist. This barrier keeps the acid from biting the metal, and therefore it's more efficient to suspend a metal piece upside down (figure A) or stand it on its side in the acid bath. This positioning allows the residue to fall off the metal, leaving a clean surface for the acid to bite. To speed up the process of the ferric-based acid etching, it's helpful to keep the solution moving. You can use a fish tank pump, a chemical magnet stirrer (figure B), or a spray-etching machine to move the solution. (You can even place the container of acid on top of a running clothes dryer!) Heating the acid speeds up the etching even more, but it creates more fumes, and I don't find it particularly necessary.

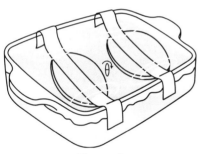

Figure A

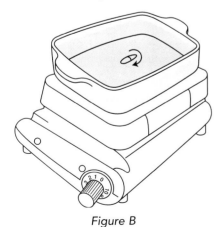

Figure B

ETCHING TIME

The length of time you leave a piece in an acid bath depends on the thickness of the lines in the resist, the strength of the acid, and how deeply you want to etch the piece. Use a needle tool or tweezers to frequently check the etched depth. Always wear rubber gloves when you do this, and make sure the gloves are not porous so the acid won't come in contact with your skin. The strength and the age of the acid you use effects the amount of time it takes to etch the metal. Begin checking the metal after approximately 30 minutes, and then depending on what you find, continue checking at intervals appropriate to the etching depth you see. Weak acid can take several hours or even overnight to etch metal.

CLEANING ETCHED METAL

When cleaning a piece etched in ferric chloride, you must neutralize the mustard-colored residue left on the metal surface. Use a solution of ammonia and water and scrub the metal with a toothbrush or scrub brush until all the residue is removed. (Be sure to use ammonia in a well-ventilated area!) If you heat the metal with the mustard-colored residue left on its surface, you'll create a toxic chlorine gas that you don't want to breathe!

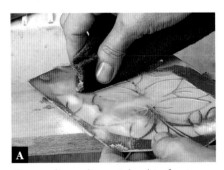

On sterling silver etched in ferric nitrate, the residue looks gray, and it's not as toxic (see photo A). Use a paste of baking soda and water to clean and neutralize these pieces. Don't let the residue dry on the metal before you scrub it off, or the surface will be very difficult to

Above: Keith A. Lewis *Charm*, 2002. 7¾ x 1 x ¼ in. (20 x 2.5 x .25 cm). Sterling silver, 24-karat gold plate, enamel, horn, bone. Photo by Doug Yaple.

Far right: Linda Darty *Backyard Bridges*, 1997. 6 x 4 x 4 in. (15.2 x 10.2 x 10.2 cm). Sterling silver, enamel. Photo by artist.

clean. If you can't clean the metal right away, simply put the piece underwater, and then later soak and clean it in the appropriate solution to neutralize the acid.

ACID STORAGE & DISPOSAL

Store ferric chloride and ferric nitrate in opaque containers to preserve their shelf life. Label each bottle with the date you mixed the solution, and note on the container how many times and how long the acid has been used. You can reuse both ferric chloride and ferric nitrate, but the longer the acids are used and the more pieces they etch, the weaker the acids become. Weaker acids etch metal more slowly, and ferric nitrate seems to weaken more quickly than ferric chloride. Etching slowly is actually preferable for some purposes, because slower acids etch more cleanly with fewer undercuts. When an acid no longer etches metal in a reasonable amount of time, contact your local hazardous waste authorities regarding the acid's proper disposal. The ferric ions alone are not particularly hazardous to the environment, but what is left in the solution after etching can be very hazardous. In solution, acids become a combination of the original chemical with copper chloride (ferric chloride) and silver nitrate (ferric nitrate).

Never pour used ferric chloride solution down a sink drain because of the residual copper ions left in it. To make it safe for disposal, neutralize the acid with sodium carbonate (washing soda) or sodium hydroxide until the pH value rises to 7.0 or 8.0. (Test pH levels with indicator paper.) Copper is left in an acid solution as sludge. Allow the sludge to settle, pour off the liquid, further dilute the liquid with water, and then you can pour it down the drain. Only use this method if your studio or household drains are connected to a sanitary sewer system with a water treatment plant. Collect the sludge in plastic bags and dispose of it as required by your local waste authority. (Or, you can simply have them pick up the entire container of spent acid if you don't want to go through the separation process.)

You can remove the silver dissolved in the ferric nitrate solution by adding sodium chloride solution to the mix. The silver will precipitate out as a white solid silver chloride. You can dilute and dispose of the resulting solution without concern. But most often, even when you're using ferric nitrate,

you're etching sterling silver that does contain copper or perhaps you're even etching copper. This means that copper ions are dissolved in the solution, and you should dispose of the ferric nitrate acid solution as you would a ferric chloride solution. The easiest way for me to remember it is to think of the quote: "If you can't drink it, don't put it down the drain!" I simply call hazardous waste authorities to pick up my used solutions of ferric-based acids.

RESISTS FOR ETCHING

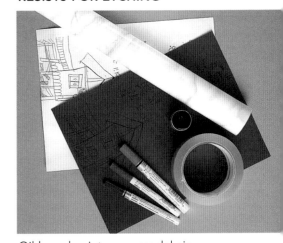

Oil-based paint pens, model airplane paint, heat transfer paper, clear packing tape, and contact paper can be used as resist materials for etching with ferric-based acids.

Top: Tamar De-Vries Winter
Mezuzah, 2001. 3³⁄₁₀ x ⁹⁄₁₀ in. (8.4 x
2.3 cm). Sterling silver, gold foil,
enamel. Photo by James Austen.

Bottom: Deborah Lozier *Chess
Pieces: King & Queen*, 2003. 4 x 1½ x
1½ in. (10.2 x 3.8 x 3.8 cm). Copper,
enamel, patina. Photo by artist.

Oil-based Paint Pens & Model Airplane Paint

These materials are easy to apply and work great in the ferric-based acids. Both oil-based paint pens and oil-based model airplane paint will resist ferric nitrate or ferric chloride long enough to etch half way through 14-gauge metal. Apply them thickly! (It is necessary to check the piece during etching and reapply the paint pen in places where it may have lifted. Keep paint pens handy for touch up with any resist.) If you do see an area that has lifted and needs another application, don't touch the surface and risk losing other areas of resist; simply dry the piece with a hairdryer, then apply more resist pen.

Heat Transfer Paper

This is a very stable and non-toxic method for transferring an exact duplicate of an intricate design onto a piece of metal. Use a photocopier or a computer with a laser printer to transfer a black-and-white image onto the dull side of the blue transfer paper. Then use the heat of a simple household iron (or some other flat heating device, such as a photo-mount press) to transfer the image from the paper to the metal. Step-by-step instructions for using transfer paper are included with the Etched & Enameled Brooch project on page 146.

Additional Etching Resists

Packing tape, contact paper, and clear electrical tape all hold up as resists in ferric-based acids. You can use them to create straight sharp lines for an etched surface. These materials also work especially well to prevent the back side of the metal from being "bitten" by the acid.

ETCHING DEMONSTRATION

1. Use colored pencils to design the piece you wish to make.

2. Use tracing paper coated with waxy colored pencils to transfer the drawing (see page 120).

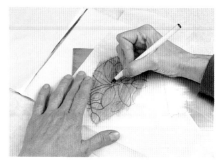

3. Draw over the transferred lines with an oil-based paint pen to create the resist.

4. Apply packing tape to the back side of the metal piece as an etching resist.

5. Carefully immerse the metal in an acid bath. The amount of time you leave the metal in the acid varies depending on the strength and age of the solution you're using.

6. The acid etches more quickly if it's moving. Acid should be placed in a glass or plastic container. Here I am using a chemical stirrer to keep the acid moving. (A magnet placed in the solution spins when the device is turned on.)

7. Check the piece frequently.

8. Keep an oil-based paint pen handy to touch up the resist as needed. Before applying any additional paint pen to the metal surface, rinse and blow-dry any areas where the resist has lifted.

The sample above shows the completed champlevé etching. The finished enameled candleholder made from this plate is shown on page 7.

FAST FACTS & HELPFUL HINTS

— Wear impervious rubber gloves, aprons, and eye protection when using acids.

— Work with acids in a well-ventilated area. Do not breathe fumes, and work under a fume hood if possible.

— Keep baking soda and ammonia nearby to neutralize any spills.

— Always keep acids covered when they're not in use.

— To reduce the chance of acid splashing on you, always add acid to water, not water to acid.

— Label all stored acid with the solution and the date. To help you determine its strength for estimating etching times, keep notes on how often the acid has been used.

— Be sure the metal is absolutely clean before you apply a resist.

— Always test a resist on a small metal piece before you attempt to etch a major piece.

— Frequently check the progress of an etching and use an oil-based paint pen as needed to touch up resists that break down.

— Clean and neutralize all etched pieces after removing them from acid. Once this is done, you can form and prepare the metal for enameling.

Linda Darty *And a River Runs Through It*, 1982. 4½ x 8 in. (11.4 x 20.3 cm). Sterling silver, enamel. Photo by Dan Bailey.

Transferring a Drawing to Metal Before Applying a Resist

Your design can be transferred to the metal, prior to applying the resist, by using graphite or dressmaker's paper. Place the transfer paper on the metal, position your drawing on top, and lightly retrace it. You can also make your own transfer paper as demonstrated on page 118). You only need a ghost line to follow with the resist pen.

ADDITIONAL CHAMPLEVÉ TECHNIQUES

Kristen Anderson taught me the saw and solder technique, which she learned while she was in Norway working at the David Andersen enameling studio. (A photo of her piece is on page 116.) If you know how to saw, you can cut designs out of a thinner gauge metal and solder them to a heavier base metal. Kristen uses 22-gauge sterling silver for the top sheet and 18-gauge sterling for the back sheet. The two sheets are sweat-soldered together very carefully so that the solder does not flow into the areas where the enamel will be inlaid. Use hard solder for joining the two pieces, not IT or Eutectic. Though it takes some sawing skill, the technique is

really quite simple. The two pieces of metal soldered together will create structural strength and if the piece is relatively small (jewelry scale) and minimally formed, counter enameling is usually not necessary. The advantage of this is that once the piece is completely enameled and finished, findings can be soldered to the bare metal on the back of the piece using hard solder (while it is turned upside down on a trivet so the surface of the enamel will not be damaged). Hard soldering essentially re-flows or re-fires the enamel. If medium or easy solder is used, the enamel might get just hot enough to pit, bubble, or crack. (Note: If you choose fine silver or copper—softer metals—for the saw and solder technique, you will need to counter enamel the piece to prevent warping and cracking. The Champlevé Brooch project on page 163 explains this

technique in greater detail.) Many 22-gauge pierced metal components (photo A) were soldered onto an 18-gauge sterling silver sheet before it was formed and made into the teapot pictured above. The fish were attached with hard solder, and the teapot was constructed with hard solder. The spout was made with IT solder and affixed to the teapot with hard solder. The piece was enameled after being completely constructed.

ENGRAVING

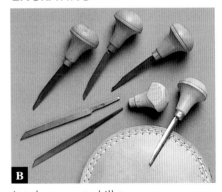

It takes great skill to engrave metal carefully enough to create even depressions for enameling. Engravers use sharp graver tools (photo B) and cut away sections of metal to form carved recesses with straight, level walls. Because the cut-away areas are so bright and

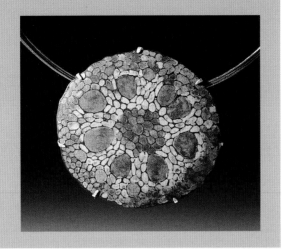

Melissa Huff *Iris Root Mandala*, 1998. 1⅞ x 2 x ⅜ in. (4.8 x 5 x .9 cm). Copper, fine silver, sterling silver, enamel; etched, champlevé. Photo by Wilmer Zehr.

reflective, transparent colors are spectacular when inlaid over the engraved surfaces.

Joan MacKarell created the engraving in this photo using 18-gauge sterling silver. The disk was first domed and the design scribed onto the metal, before it was mounted into setter's wax on a wooden block, and then supported in a sandbag while cutting. Joan suggests that you begin by cutting away from the outline towards the center of the area to be recessed. The cuts should slightly overlap, and you should work methodically around the outline, then concentrically towards the center. Once the recessed area has had a complete layer of metal removed, cut around the edge to clean it up. Repeat until the required recessed depth has been achieved, usually at least 0.3 mm. The background can be textured, and the sterling should be depletion gilded prior to enameling.

ELECTROFORMING

C

D

These pieces by Melissa Huff show electroforming. This process builds up rather than removes metal areas or lines for enameling, making it the opposite of most champlevé techniques. To electroform for champlevé, you'll apply a resist to a copper object in the areas you want to enamel. Coat the rest of the copper surface, which will be raised in the finished design, with a metal-based con-

ductive paint. Next, suspend both the object and another piece of copper in a chemical bath hooked up to an electrical circuit. A rectifier runs current through the system, and the metal from the copper is attracted to the conductive paint, depositing itself onto the painted surface (photo C). When a sufficient amount of copper is deposited on the surface, clean the piece and inlay the enamel into the resulting copper recesses (photo D).

CASTING

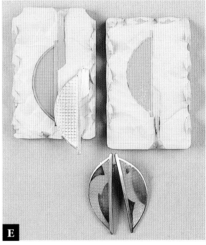

E

With the intention of enameling in the low areas of the design, you can create a wax model that includes raised and recessed areas. Once cast, you can enamel the recesses (photo E). Enamellists who make multiples for a production line will sometimes create an original piece in metal, and then make a rubber mold from it. This allows them to easily make a wax model of the original design. If your studio isn't equipped to handle these techniques, commercial casting companies will make the rubber molds, the wax models, and the castings for a very reasonable price.

METAL CLAY

Metal clay is made of fine particles of pure silver, water, and an organic binder that burns out when heated in a kiln. This material feels similar to clay, and you can shape and texture it with your fingers or with tools. Fast-firing metal clays are denser than regular metal clay, and therefore better for enameling. To use the champlevé technique with metal clay, press a tool or an embossed plate into the clay surface to create depressions. Mary Reynolds created these samples to demonstrate how she uses metal clay for champlevé. You can make a stamping tool for embossing metal clay out of basswood, plastic, or plaster (see photo A, left). Make sure to cut all the shapes in the stamp design at a right angle so there won't be undercuts in the areas you plan to enamel. Cover the edges of the stamp with copper foil tape to create a smooth finish.

Shape the piece and let it dry (photo A, center). Place the metal clay in an electric kiln and heat it until the metal particles in the clay fuse (photo A, right). This time will vary depending on the type of metal clay you use. The fired surface is somewhat porous, so to prepare it for enameling, burnish the piece with a steel burnisher or tumble it in a tumbler with steel shot. This also shines the metal, an important step if you want to inlay transparent enamel into the depressions. Inlay enamel colors into the metal clay and fire and finish the piece as you would a traditional champlevé (photo B).

CHASING & REPOUSSE

You can use the same chasing and repousse technique described in Basse Taille Enameling on pages 76–77 for champlevé enameling. When using this technique for champlevé enameling, place the metal in pitch, and use small steel tools to push back (chase) the surface you want to enamel, creating depressions to hold the glass. Turn the metal over in the pitch and push it out from the opposite side (repousse) to refine the design if you wish. If you create depressions that are too deep, the enamel might crack. If this happens, apply an additional coat of counter enamel to help equalize the expansion and contraction of the metal and glass.

ENAMELING A CHAMPLEVÉ PIECE

1. Use any technique you desire to create recesses in a metal surface. Clean the metal well. If you're using sterling silver and plan to work with transparent colors, depletion gild the metal (see page 26).

2. Prepare the enamels, sifting or washing colors as needed. Mix the enamel in a paint tray with a few drops of water and holding agent if necessary.

3. In most cases, apply a counter enamel to the back of the champleve piece. This may not be necessary if you've made a piece using the saw and solder process, or if your piece is formed from very thick metal and only has shallow depressions.

4. Wet-inlay or paint the first coat of enamel into the recessed areas. (If you're making a champlevé enamel on sterling silver, cover the entire front surface of the metal with enamel on the first firing before counter enameling, or you'll risk the formation of firescale on all uncoated areas. Because of the low acid resistance of lead-bearing enamels, you won't want to pickle a piece to clean off firescale if lead-bearing colors were used.) If desired, you can add low cloisonné wire, foils, or underglazes to a champlevé piece after firing the first enamel layer.

5. Continue to inlay enamel until the glass reaches the level of the raised metal plane. If you want a smooth surface on the finished piece, grind it with an alundum stone or use diamond-sanding sticks until the metal and enamel are flush and even.

6. If you ground or sanded the piece to make it flush, you must now sand the metal to remove all scratch marks. I suggest using carborundum paper, emery paper, or diamond sanding sticks under running water. The sanding paper will remove less enamel than the diamond sticks, so as soon as the surface is level and you like the enamel color in place, you may want to switch to paper. Only use the sanding materials in a back-and-forth direction.If you sand in different directions you create more scratches that you'll have to sand out later. Focus on making the metal beautiful. Don't worry much about what the enamel surface looks like unless you don't plan to re-fire the piece. In that case, continue working on the glass and metal using the hand-polishing steps described on pages 52 and 53.

Note:

If you're working on sterling silver, you may see firescale appear as you sand the piece. Don't worry. Once the metal looks nice you can fire it a couple of times to bring the firescale to the surface, and put it in a cold pickle bath for a short time between firings. (Most cool colors will tolerate a few minutes in pickle even if they're not completely acid-resistant. Enamels with very low acid resistance will become cloudy or matt after firing if left in the pickle too long.) Use a buffer or continue sanding to remove firescale if necessary.

7. Once the metal is finished the way you want it, clean the piece with a glass brush and refire. Be sure the metal finish is perfect

because if you refire the piece and then decide to resand it, you will damage the glossy enameled finish. The piece could also be matte finished as described on pages 45 and 52.

8. At this stage, you can oxidize the piece in liver of sulfur, brass-brush it, or polish it with steel wool or on a buffing wheel. The piece can be waxed or finished in any way that won't damage the glass. You can texture the surface of the raised metal planes, but be careful not to break or chip the glass. I've known artists to gently engrave texture on the raised planes or mask off the enamel and sandblast the metal.

FAST & FACTS HELPFUL HINTS

— If the depth of the recesses in the champlevé piece is shallow, be sure to apply the colors that you will eventually want in the piece during the very first firing. Do not apply warm colors directly on fine or sterling silver unless you've tested them! I usually apply clear flux in places where warm colors will eventually go. In most cases, you will not be wet inlaying colors as many times with champlevé as you do with cloisonné, so apply the colors you like early, and practice shading colors horizontally, blending across the piece, rather than in many vertical layers.

— If you decide you want to use cloisonné wires or granules in the piece, do it as soon as you can after the first firing. Again, this is because you may not have room for many layers of enamel.

— When stoning or finishing a champlevé piece to make it level, concentrate on the metal surface. Keep sanding in the same direction so you'll have less scratches to remove.

— If you reach the top level of a metal plane and decide, even after stoning the surface level, that you wish you could change a color or add a little more color, you can use diamond grinding burrs to alter the glass surface. Put water on the piece and work gently in the area you wish to grind out. Pick the type of burr that fits the space you want to grind. I generally prefer burrs with a rounded end. If you're using sterling silver, be careful that you don't grind down to its surface, losing the fine silver layer that you so carefully built up before enameling. If you do, you risk getting firescale in the enamel.

— If you decide early in the wet inlay process that you don't like a color, you can apply foil over the area. If you wait too late to add the foil, you might end up stoning it away during the finishing process.

Plique-à-jour

THIS BEAUTIFUL TECHNIQUE uses transparent enamels without metal backing, letting light intensify and reflect the colors like a miniature stained glass window. Good plique-à-jour work is stunning, both in the crystalline clarity of the glass and in the detail of the filigree metalwork that surrounds it. The term *plique-à-jour*, partly Italian and partly French in derivation, literally means "similar to a membrane (*plique*) stretched in a way that the light of day (*à jour*) may pass through." The enameled object should be designed so that it can be viewed or worn in such a way that light can penetrate it from either side. Bowls, goblets, hair combs, flatware, and earrings might function this way, but brooches and pendants that are worn against the body would not.

It takes patience and practice to learn plique-à-jour enameling, but it's not as difficult as it may seem. The size of the openings and the relationship of enameled areas to metal areas must be carefully planned to minimize the chances of the enamel cracking. If there is more metal than enamel, the pressure of the metal against the glass could cause it to crack. If the openings are hard edged and angular, the pressure of the two sides of metal in the angle, espe-

cially near the corner, can press on the glass and cause a crack as well. For this reason, most plique-à-jour work is designed with flowing curvilinear lines.

There are different ways to design the framework that holds the glass and different ways to apply and fire the enamel in the frameworks. Some plique-à-jour enamellists use only very cleanly washed 60- or 80-mesh enamel, so that the colors created by the larger grains will be as clear as possible. Other artists find that well washed 200-mesh enamel is easier to apply when working on curved dimensional surfaces. If you're a beginner, start on a small, flat piece, trying different methods for constructing and filling the piece until you find the one that works best for you.

CREATING THE PLIQUE-À-JOUR FRAMEWORK

Here are four methods enamellists use to create the metal framework.

Pierced Plique-à-Jour

This method involves using a jeweler's saw to pierce small, carefully drawn shapes out of a piece of 16- or 18-gauge fine silver, fine gold, or copper. (Copper would require pickling between firings to clean firescale.) The heavier 16-gauge

metal will have taller walls and allow you more room to fill with color.

To begin, drill a small hole in each of the shapes to be removed so that you will have a place to insert your saw blade. Thread the saw blade into each hole and saw around the shape (photo A), leaving the metal between the shapes intact, and letting the metal shape (which will be left as openings for enamel) drop out. Saw out the holes in the center of the design first, working your way to the outside so the metal is well supported by the excess around the edge. Do not saw off excess metal from the outside of the design until all the cells have been sawn out (photo B); this gives you something to hold on

HISTORICAL HIGHLIGHT

In *The Treatises of Benvenuto Cellini on Goldsmithing and Sculpture*, the renowned Italian goldsmith tells a story about a "wonderful and priceless work" shown to him in 1541 by King Francis in Paris, France. He describes seeing, "a work most rare, a work such as may perchance never again be executed…" When asked to explain how it was made Cellini states that, "all the noble assembly that waited on his Majesty thronged around me. The King declared he had never seen work of so wondrous a kind…" and Cellini explained to them that gold filigree wire was formed into designs inside an iron bowl, painted with gum tragacanth to hold them in place, and then enameled. (A clay slip was painted inside the iron bowl prior to lying in the wires.) He further explained that the filigree work is "as good one way as the other, soldered or not soldered." After inlaying the enamel he states that "to begin with you must only subject it to a slight heat after which, when you have filled up any little openings with a second coat of enamel you may put it in again under a rather bigger fire, and if it appears after this that there are still crannies to be filled up, you put it to as strong a fire as the craft allows and as your enamels will bear."

Gustav Guadarnack for David Andersen, Norway *Plique-à-jour Goblet,* early 20th century. 7 x 3½ in. (17.8 x 8.9 cm). Fine silver, enamel; plique-à-jour.

A strong fire it must have been indeed, to fuse the glass in this primitive furnace inside an iron bowl!

Today, the most familiar plique-à-jour is of the Art Nouveau style. It was executed at the turn of the twentieth century by artists such as Gustav Gaudernack, who was appointed permanent designer at Norwegian David Andersen's goldsmithing firm in 1892. Dating from 1901–1908, the works in plique a jour enamel by Gaudernack are among Norway's greatest treasures and important contributions to the European art nouveau style. Norwegian plique-à-jour enamel was not made using a temporary backing. Instead the cells were formed from flat wire and the enamel was inlaid using the surface-tension technique.

Bob Corson is a collector of plique-à-jour enameled objects dating between 1890 and 1910. Corson explained in *Glass on Metal* magazine that during the gay 90s, when the newly rich were off to see the world, they often purchased souvenir enameled spoons while traveling. Collectible spoons were available in handcrafted silver, simple or perhaps engraved or enameled with a picture of a local scene. Sometimes only the stem or finials were decorated, but the most expensive of these spoons included plique-à-jour enamel in the spoon bowl as well. The wirework was elegant with the beautiful transparent colors inlayed in art nouveau designs. At the turn of the twentieth century the art nouveau period was taken over by Art Deco, and plique-à-jour souvenirs spoons disappeared almost completely in the decade before World War I.

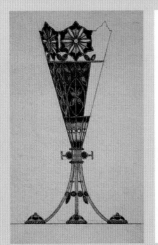

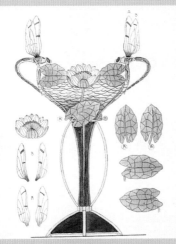

Gustav Guadarnack for David Andersen, Norway. *Designs for Plique-à-jour Pieces*, early 20th century. Ink and watercolor on paper.

Three Magnum Plique-à-jour Souvenir Spoons, Norway, circa 1890–1910. Makers (left to right): Marius Hammer, Bergen; David Andersen, Kristiana; Unmarked, possibly J. Tostrup, Kristiana. Courtesy of Bob Corson.

to while you saw. Keep the saw-blade perfectly straight up and down so that the walls are clean and will need no filing. If you do need to file the edges after sawing because they're rough or cut at an angle, keep the file perpendicular and file slowly. The burr that is left by the sawing process can actually help hold the enamel in place during the wet inlay process, so do not file it off unless it's necessary.

The piece can be enameled using the capillary action fill method, the mica backing technique, or thin metal foil can be burnished to the back of the pierced piece, then peeled away after enameling. (These inlay methods are explained on pages 127 and 128.)

Soldered Filigree

Valeri Timofeev, an artist from Latvia, researched and re-discovered many technical secrets of plique-à-jour enameling that were lost during the Russian Revolution. He began by replicating the work of Russian masters such as Fabergé, and eventually developed his own soldered filigree method, which is briefly described here.

First, fine silver or fine gold wire is glued together over a steel form and bound with steel wire Next, eutectic solder made of pure copper and fine silver is filed into shavings that are sprinkled over the piece before it is soldered in sections. After soldering, the piece is heated and pickled (using depletion gilding, described on pages 26 and 27) to remove the copper firescale from the solder. The piece is burnished after pickling, and the process is repeated until the silver is powdery white. The piece is not burnished after the final pickling. The frosty white piece is enameled using the capillary action fill method described on page 127.

Soldered filigree vessel by Valeri Timofeev, before and after enameling

Cloisonné Wire & Acid Etching (Shotai' jippo)

This method is similar to cloisonné enameling except that one side of the copper piece is left uncoated with enamel and later etched off with acid.

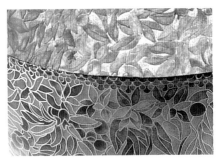

Detail of a plique-à-jour bowl by Fay Rooke, made with the cloisonné wire acid etching method

To begin, fire three transparent coats of enamel to a copper form and do not apply counter enamel to the back. (Coating the back side of the copper with a firescale retardant like ball clay or Scalex will alleviate the need to pickle between firings.) Create the design for the plique-à-jour using cloisonné wire, and apply it to the enameled surface. Design a border with smaller openings around the top edge of the piece in such a way that every wire touches another wire to make a rim. (This helps prevent cracks in this fragile area.)

In several gentle firings, fill the cloisonné partitions as evenly as possible with clean transparent colors; then grind and finish the filled surface as you would any cloisonné piece.

Coat the finished enamel with a lacquer varnish that resists the effects of acid. (You may want to coat a small edge of the back piece of copper as well, so that a rim or

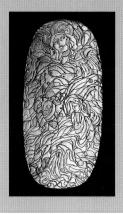

edge is left around the piece for support.) Submerse the piece in an acid solution until the copper backing etches away. After every trace of copper is removed (it might be necessary to use wet diamond burrs to remove the last bits), clean the varnish off the piece. It can then be cleaned with a glass brush or carefully buffed.

Cloisonné Wires on Mica

This is a simple way to make a small plique-à-jour enamel if you have little metalworking experience. The samples in these photos were made by Whitney Boone to demonstrate this technique.

Use eutectic solder to construct a fine silver framework from heavy round or square wire. (The frame should be about as tall as the cloisonné wire that will be used in its interior.)

Within this frame, form cloisonné wire into small shapes so that they're all touching each other for support.

Using a piece of mica as a temporary backing, inlay enamel as you would for any cloisonné piece (this will take several firings).

Lift the piece from the mica when complete and carefully use wet emery paper or diamond sanding papers to remove traces of the mica from the back.

ENAMELING THE PLIQUE-À-JOUR FRAMEWORK

Mica, as described above, can be used to support the back of the plique-à-jour piece during enameling, but it's only appropriate for small, flat pieces. Some of the mica does adhere to the glass after it's enameled, so this technique requires careful sanding to remove the residue. Following are two other methods that are typically used for inlaying enamel into a plique-a-jour framework.

Capillary Action Fill

Use this method for pieces with no temporary backing, such as pierced plique-à-jour or soldered filigree. Detailed instructions for making

earrings using this technique are featured on page 167.

1. Wash transparent enamel colors, such as 80-mesh to 150-mesh, very well. Mix a small amount of a holding agent in water (approximately one part binder to five parts water) and use this mixture to saturate the enamel powder. To minimize air bubbles, let the mixture settle into the powder rather than stirring it.

2. Use a spatula or paintbrush to fill the cells in the filigree piece with wet enamel. If the cells are small and the consistency of the enamel is correct, surface tension will hold the color in place.

3. Use a cotton or linen lint-free towel to wick out some of the liquid as needed. As you inlay, gently tap or vibrate the piece to smooth the enamel surface.

4. Fire the inlaid piece on a trivet to the orange-peel stage. [I usually fire at approximately 1430° F (776° C) for about 45 seconds.] The

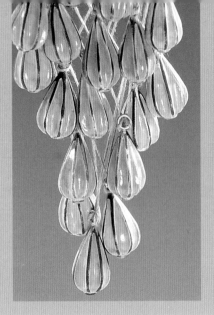

Above: Valeri Timofeev *Bowl,* 2001. 3 x 6½ x 6½ in. (7.6 x 16.5 x 16.5 cm). Silver, enamel; plique-à-jour, gilded. Photo by artist.

Left: Diane Echnoz Almeyda *Water Lillies,* 1999. 6 x 4¼ x 3½ in. (15.2 x 10.8 x 8.9 cm). Fine silver, enamel plique-à-jour. Photo by Dan Loffler.

Far left: Tzu-Ju Chen *Pendant,* 2000. 3⅕ x 4 in. (8.1 x 10.2 cm) pendant; 15 in. (38.1 cm) necklace. Sterling silver, enamel, silk; plique-á-jour

enamel will fire as a fine web over the structure, with multiple holes. Turn the piece over, and then fill and fire it again. This application fills the holes with an even distribution of enamel that is two or three grains thick.

5. Repeat steps 2–4 to inlay and fire additional enamel coats until the filled cell openings are concave on both sides, but have no holes. (The cells should be filled, but the enamel should not come up over the wires.) Each time you repeat these steps, reverse the side you inlay with enamel and let the enamel completely dry before each firing.

6. Stone or sand the surface as needed to remove any enamel that coats the wires. If any glass cracks, apply a thin coat of enamel and re-fire the piece. Finish the final surface with diamond sanding paper or wet carborundum paper, or re-fire the glass to a glossy surface.

Burnished Copper Foil

1. Cut thin copper foil to fit the back side of a pierced or soldered plique-à-jour framework.

2. Cut notches in the foil edge so you can push it up and over the sides of the framework.

3. After burnishing the back with a blunt tool, add color in layers until the top surface is level. Carefully sand the front side of the framework.

4. Gently peel the copper foil away from the back side of the enameled piece. If you paint an asphaltum or a lacquer resist on the front of the piece, you could etch the foil off the back with nitric acid or ferric chloride.

Additional Techniques

Right: Rebekah Laskin *Brooch*, 2000. 2 x 2½ in. (5 x 6.4 cm). Copper, sterling silver, enamel, epoxy resin. Photo by artist.

Far right: Ken Rockwell *Brooch*, 1979. 2½ x ¾ in. (6.4 x 1.9 cm). Sterling silver, copper, enamel. Photo by the artist.

NOW THAT YOU UNDERSTAND a number of different ways to work with glass on metal, you should take that knowledge and see what else you can do with it! Experiment! Use what you've learned with an open mind, trying new things, making samples, and pushing the material beyond what you know it will do. I remember taking a workshop with Jim Malenda years ago, and sometimes, when a student asked him how a color might look when applied in a certain way, he replied "Don't know, might work, make a test." When the class ended, we gave him a T-shirt with those words on the front, and I chuckle now when I find myself saying the same thing to my own students. It's simple, fun, and inexpensive to make tests and experiment with enamel. Relax and have fun in your studio. Enjoy each piece—letting your successes and also your mistakes—lead you to try new things.

WORKING WITH GLASS LUMPS, THREADS, & BEADS

Glass lumps, seed beads, and milliefiore beads can be placed on enameled surfaces and fired so they fuse into a base coat of enamel. Available from enamel suppliers, opaque enamel threads can be used to create special effects. The samples on this page were made with some of these individual materials to give you an idea of how they appear on a fired enameled surface. Depending on how hot you fire your piece, the lines, beads, and lumps will bleed and flow smoothly or remain raised. The threads and lumps can be gently laid into a sifted coat of enamel with tweezers before the sifted coat is fired. If the base coat has already been fired and is slick, you might need to dip these items into a holding agent before applying them to the piece. By using these materials thoughtfully and with restraint, you can add interesting details and texture to a finished piece.

Glass threads were fired on all three samples in photo A. The sample on the left incorporates the threads exactly as they were purchased. The center sample used the same type of commercial glass threads, but they were crushed and sprinkled on the piece prior to firing. The piece on the right was made by sifting crushed threads through an 80-mesh sifter. As a final step, all of the samples were given a matte finish with a chemical glass etching cream.

In the sample in photo B, opaque lumps of enamel were placed on top of a black base coat. The sample was fired until the lumps smoothly fused into the glass.

To create the samples in photo C, Annie Grimes secured tiny seed beads in place with a holding agent, and then the pieces were fired into an enameled surface.

"In art, there is inevitably a seepage from your life. I believe that what happens in my work is there because it wanted to happen."

Dorothy Sturm

"Often there is tragedy, but sometimes something special can occur. I think perfection is nice, but I've never achieved it and I've never grown tired of the exciting gamble."

June Schwarcz

but only after making careful notes regarding what had happened so she could use the information as she continued her research.

During the same time period, June Schwarcz was in a different part of the country, enameling on simple spun vessels. As she became more comfortable with the material, she too developed ways of working with vitreous enamel that were experimental and expressive. Inspired by the California landscape, Schwarcz formed pieces with great energy and gesture, and they were rich with the textures and colors of her surroundings. She used whatever techniques were appropriate for her ideas, including basse taille, champleve, cloisonné and plique-à-jour. To create textured surfaces, she experimented with adding materials to enamel, such as sand, clay, and slip. She used electroforming techniques to build up copper, and then surrounded it with smooth glossy enamel. June Schwarcz continues to inspire generations of enamellists who marvel at her inventiveness with glass on metal.

Dorothy Sturm Untitled, circa 1960. 12 x 12 in. (30.5 x 30.5 cm). Copper, enamel, glass. Collection of Mr. and Mrs. Downing Pryor. Photo by Robert Davis Carrier.

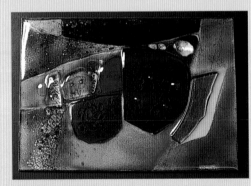

June Schwarcz Untitled, 2003. 6½ x 5 in. (16.5 x 12.7 cm). Copper, enamel; electroformed, iron plated. Photo by Lee Fatherree.

June Schwarcz and Dorothy Sturm began enameling in the early 1950s. Both women were influential in developing and working with experimental enameling techniques.

Dorothy Sturm worked in Memphis, Tennessee. She used large shards or small bits of transparent and opaque glass, and slumped and fused these pieces to enameled surfaces in a collage. She used glass from everywhere: Vienna, Venice, or simple crushed-up old soft drink bottles. She occasionally added beads or bits of screen or wire to the pieces, but mostly she worked with pure glass. The colors expanded and changed when fired and Sturm responded to those changes and learned from them. She studied the science of combining glass and metal, and gained control of the process so that very little was left to chance. When she was not pleased with a fired piece, Sturm would destroy it,

Dorothy Sturm Untitled, circa 1968. 12 x 12 in. (30.5 x 30.5 cm). Copper, glass, enamel. Photo by Robert Davis Carrier.

June Schwarcz Untitled, 2003. 11½ x 5 in. (29.2 x 12.7 cm). Copper, enamel; electroformed. Photo by Lee Fatherree.

Making Glass Chips

You can use enamel lumps and threads to make multicolored glass chips.

1. Sift a coat of enamel on one side of a thin long strip of copper.

2. Apply or sift threads, lumps, and patterns of enamel onto this strip and fire the piece.

3. Bend the strip and multi-colored chips will flake off the copper surface. Store these chips for use on enameled surfaces for special effects.

Another method of creating colorful glass shards is to sift and fire multiple layers of opaque enamel onto copper, and then stone and grind the surface to reveal a pattern. The following samples, created by Barbara Minor, show a method for creating a mosaic pattern using these enamel shards.

Photo A: a black counter enamel and a black base coat were applied to the metal.

Photo B: the enamel shards were created by sifting multiple layers of opaque enamel onto a piece of copper that hadn't been counter enameled. After firing, the copper was bent, and the enamel shards were broken off the surface.

Photo C: a holding agent was used to secure the enamel shards to the black enameled metal. The piece

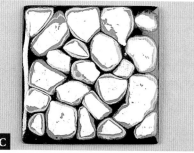

was fired until its surface was smooth. (This step can take multiple, extended firings to get the thick enamel shards to fuse and smooth. You may need to fill deep indentations between shards with additional black enamel.)

The smooth surface was ground down by hand with an alundum stone. (You could also use lapidary grinding equipment, such as a rubber drum sander.) It may be necessary to re-fire the piece during the grinding process to "heal" cracks and further smooth the surface.

Photo D: this completed sample has a matte finish.

Kathleen Browne *Double Trouble*, 2002. 1⅞ x 4¾ x ½ in. (4.8 x 12 x 1.3 cm). Fine silver, sterling silver, enamel. Photo by artist.

ENAMELING ON COPPER FOIL

It is especially fun for beginning metalworkers to use copper foil because it can be easily manipulated and shaped with the hands or simple tools. Try roll milling the foil with an embossing plate to create texture, hammering it, folding it, or simply drawing on it. Even very young students can draw with a plastic pen top or wooden stylus on copper foil that is placed on a magazine or some other soft backing. The embossed fine drawings and textures look beautiful under glass. Enamel collects in the low lines, or around the raised areas and it pools, to create dark and light values. If the pieces are flat, you should fold the edges under for structural strength. Clean the metal well before enameling. You might want to apply dark underglaze pigment to the low recesses before covering the piece with a transparent sifted or liquid color. Liquid white also creates an interesting surface as it turns green and brown in the areas where it covers the piece more thinly. You can use your fingers to wipe the dry liquid enamel off the high spots if you wish, before firing. Since the metal is so thin, the firing time will be shorter than it is when you use heavier gauge metal.

Each sample in the collection above was made by Dindy Reich on embossed copper foil. The colors used were liquid white and liquid clear enamel.

DECALS

Printed imagery has been used on enamels since the 18th century. Today, you can purchase solid, patterned, and pictorial decals from enamel and ceramic supply companies. You also can make decals from your own images by creating a photo silkscreen from the image, and then screening overglaze enamels onto a special decal paper. Depending on the type of overglaze enamel you use, after you screen it onto the paper and let it dry, you may need to use a squeegee to cover the entire image with an organic cover coat. After being soaked in water, this thin plastic-like covering releases from the paper so you can slide

the decal onto your piece. With some overglaze enamel pigments, however, you need not apply an organic cover coat.

If you know how to apply a decal to a car window, you can apply a decal to an enameled surface. You simply cut out a handmade or commercial decal (photo A), soak it in water until it easily lifts off the backing paper (photo B), and then slide it onto a fired enamel surface (photo C). Blot the decal once you position it on the enamel, allow it to completely dry, and then fire. You'll need to experiment with different firing temperatures for different types of decals, but always fire until the enamel surface glosses. Most decals can be fired at a slightly lower temperature than regular enamels. You can determine the firing temperature and time by the type of painting enamel used to make the decal. Some decals need to be slowly brought up to firing temperature so their overcoats will burn out before their color fires down. Some commercial pottery decals are made with very fine painting enamels. Fire these in cooler kilns or they will burn out. Decal transfers make it easy to duplicate an image from an original drawing or text, and to replicate it as many times as desired.

SILKSCREENING ENAMEL

To create stenciled images and patterns, you can make a silkscreen from a drawing and push or sift dry enamel powder through it (photo D). If you don't wish to make a silkscreen, take your original artwork to a T-shirt printing company, and they can produce one for you.

Photo E is a sample by Barbara Minor. It shows 100-mesh red enamel sifted through a photo silkscreen onto a white enamel base coat and fired.

The following samples by Kathleen Wilcox show drawing, painting, and silkscreen printing incorporated into a collaged enamel wall piece. A black-and-white drawing or transparency was used to make the photo silkscreen. Using the edge of a piece of matte board as a pusher, dry enamel was pushed through the screen onto the metal without any holding agent.

The first layer of enamel, the heron shape, was silkscreened with dry enamel powder on bare copper. Different colors of opaque enamel simultaneously were pushed through the screen. The image was fired until fused (photo F). To make the leaves, two different opaque green enamels were pushed through a different silkscreen. The piece was re-fired until the enamel fused.

A black coating of firescale oxidation built up on the exposed copper areas. To create a texture in the background, the firescale was scraped off with a grinding tool in a flexible shaft machine. Using a 100-mesh sifter, clear enamel was sifted

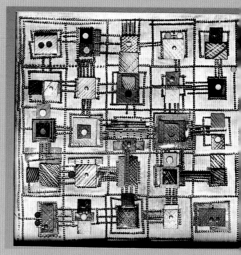

Top: **Shana Kroiz** *Mosaic Path Brooch Series*, 1999–2001. 3 x ½ in. (7.6 x 1.3 cm) each. Copper, enamel, silver, carnelians, pearls; die formed. Photo by Norman Watkins.

Center: **Elizabeth Turrell** *Dream Map*, 1996. 12 x 12 x ½ in. (30.5 x 30.5 x 1.3 cm). Copper foil, enamel; perforated, incised. Photo by artist.

Bottom: **Robert W. Ebendorf** *Fish In Hand*, 2001. 7 x 5½ in. (17.8 x 14 cm). Steel, enamel, wood. Photo by Linda Darty.

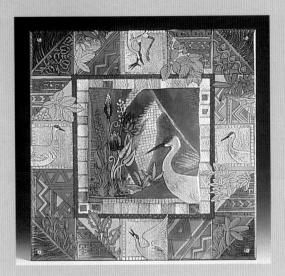

Left: Christina T. Miller *Capital: Aggravated Decay*, 2002. 7½ x 3 x 3 in. (19 x 7.6 x 7.6 cm). Sterling silver, steel, wood, hidden metronome, antique pulley, rope, enamel; rust printed, limoges, kiln fired, torch fired. Photo by Scott McMahon.

Far left: Kathleen Wilcox *Floridance Quilt: Crane Games*, 2003. 10 x 10 in. (25.4 x 25.4 cm). Gold foil, silver foil, overglazes, liquid enamel; inlay, sgraffito, silkscreened, sifted.

over the whole piece, and it was fired (page 133, photo G). The result of the clear enamel layer varies according to the time and temperature of the firing.

A crow quill pen was used with black overglaze to outline the details of the heron and the leaves. Yellow overglaze was used on the heron's beak, and the white and green areas in the piece were also touched up with overglaze (photo A).

RAKU ENAMELING

In this procedure, used often by potters, the fired piece is taken out of the kiln and immediately placed in a metal container with a combustible material (like sawdust, leaves, newspaper, pine needles, or wood shavings) to create an oxygen reduced atmosphere, thereby affecting the colors in the glass. More organic material can be thrown on

top of the piece so that it is buried if you wish. The material is allowed to flame and burn for a minute or so before a lid is clamped down on the container to trap the smoke. As the oxygen is consumed in the container, the oxygen in the glass is drawn to the surface, creating interesting effects, ranging from metallics to iridescence depending on the enamel. Shapes of the leaves or pine needles used can leave interesting shadows on the piece if they come into contact with the glass quickly enough to burst into flame. Experiment to learn which colors work best for you. Be sure to work in a well-ventilated area (preferably outdoors) and have access to water nearby.

Jean Tudor made these samples illustrating some effects from

raku firing. Sample B was coated with three different lead-free enamel colors, and then placed in a container with layers of wet newspaper. Raku firing produced swirls of copper in the green and the copper color that extends down into the smaller areas of the piece.

Sample C had gold leaf applied to its lead-free enameled surface. After firing, it was placed into a raku container with dried leaves that left interesting metallic shapes on the blue-and-green surface of the glass.

Right: Tzu-Ju Chen *Brooch*, 2000. 3 x 1¾ in. (7.6 x 4.4 cm). 18-karat yellow gold, copper, sterling silver, enamel.

Far right: Sharon Massey Untitled Rings, 2003. 2 x 1 x 1 in. (5 x 2.5 x 2.5 cm) each. Sterling silver, copper, enamel. Photo by Robert Diamante.

USING ENAMEL TO FUSE PIECES

Depending on the size and thickness of the metal and the colors you choose, you can use enamel to fuse two pieces together. If you plan to try this, I encourage you to test the colors and the thickness of the enamel you use in relation to the shape and size of the work. In some cases, pieces may crack, but you might find that this joining method works very well.

Photo D: holes were drilled in the metal piece and wires were pushed through the holes for a tight fit.

Photo E: enamel was sifted on the back side of the piece to hold the metal elements together, fired, and then sifted on the front. Sharon Massey's finished ring is pictured above.

CREATING RUST PATTERNS ON AN ENAMEL SURFACE

Rust contains iron oxide, and Christina T. Miller has experimented imprinting this substance on an enameled surface to create interesting textures. After counter-enameling a piece, fire a light-colored enamel base coat on the front side of the metal. Place a damp towel in a lidded plastic or glass container. Place the enameled piece on the towel and rest a collection of damp or wet rusty objects on top of the enamel (photo F). Cover the metal with another damp towel, close the lid on the container, and let it sit for a few days. Frequently check the piece and add more rusted metal to its surface if needed to create the desired affect. When there is an adequate amount of rust left on the surface of the enamel (photo G), dry the piece and fire it until the base coat re-flows. (If you apply the rusted metal to thin steel plates that don't have counter enamel, you can saw out and shape them after firing.)

FAST FACTS & HELPFUL HINTS

— Make a lot of small tests so you won't waste material or time on an experimental technique that may or may not be successful.

— Use inexpensive enamel colors for experiments. Reds, oranges, and yellows usually cost more than other colors.

— Frequently clean your workspace.

— Keep good notes in a journal, writing down every step so you can remember what you did.

Projects

Project Kits

There are specific enameling tools and materials you'll use for almost every project. To abbreviate the length of the supply lists, I've made four "kits" you can refer to as you prepare to enamel. Since you won't need every item from a kit for every project, I encourage you to read a project's instructions in its entirety prior to starting.

Kathryn Osgood *Mourning Locket,* 2003. 1⅜ x ⅝ in. (3.5 x 1.6 cm). Copper, silver, enamel, gemstone. Photo by Robert Diamante.

Enameling & Finishing Kit
Alundum stone
Bucket, container, or coffee filter
 and jar
Colored pencils
Cotton, linen, or lint-free brown
 paper towels
Diamond sanding sticks
Dust mask
Fiberglass brush
Green kitchen scrub pad
Holding agent (also known as a
 gum binder or gum tragacanth)
Permanent marker, fine tip
Sable paintbrushes and other
 assorted brushes
Scissors, small and sharp
Scouring powder or pumice
Sifters
Sketchbook
Small glass or plastic containers with
 tight-fitting lids for storing enamel
Soft brass brush
Spatula or small tool with a flattened
 or domed end
Tracing paper
Tweezers
Watercolor paint tray or plastic spoons
Wet/dry carborundum paper, 320,
 400, and 600 grit

Enameling Kiln Kit
Fireproof gloves
Firing fork or trowel
Firing screens
Firing trivets
Heatproof surface
Kiln that reaches at least
 1500° F (815° C)
Mica sheet

Metalworking Tool Kit
Bench pin
Burnisher or bezel pusher
Center punch
Chasing hammer
Chasing tools
Dapping block & dap
Dividers or compass
Drill bits
Flexible-shaft machine, drill press,
 or hand drill
Half-round medium-cut or fine-cut
 file, 6 inches (15.2 cm) long
Jeweler's saw frame
Mallet (wooden, rawhide, or plastic)
Needle files
Pliers (flat-nose, chain-nose, and
 round-nose)
Ring mandrel
Saw blades, 3/0 and 2/0
Scribe
Small round nose pliers
Steel block
Table shear or hand shear
Wet/dry carborundum paper,
 assorted grits

Soldering & Annealing Kit
Copper tongs
Heatproof surface
Firing screens
Firing trivets
Medium to large torch tips
Pickle (sodium bisulphate)
Pickle pot, such as a slow
cooker
Propane or acetylene
 soldering torch
Safety glasses
Sharp tweezers
Small brush for applying
 soldering flux
Solder (hard, medium,
 and easy)
Soldering flux
"Third-hand" tweezers
Water bowl for quenching
 after pickle

Stencil & Sgraffito Sifted Earrings

Jennifer Mokren

Combine sgraffito and stenciling techniques to create these dramatic sifted earrings, and then set them in handmade silver bezels.

over the holding agent, and fire until fused.

3. If you're using a transparent base coat, you may want to clean the firescale off the front of the piece in a pickle solution. If you're using an opaque color, just lightly scrub the piece under running water until the water sheets off the metal. Apply a holding agent to the front side of the clean copper. Sift on the base coat, and fire until the enamel is fused (see photo). Let the piece completely cool.

4. File or sand the edges of the metal to remove firescale. (Remove firescale from the edges of a piece

Techniques

Sifting with stencils

Sgraffito with sifting

Producing an etched matte finish

Bezel setting with sterling details

What You Need

Copper sheet, 22 gauge, or precut copper shape

Fine silver bezel wire, 26 gauge

Sterling silver sheet, 24 gauge

Sterling silver round wire, 20 gauge (for ear hook)

Opaque or transparent enamels

Glass etching cream (optional)

Sifter, 80 mesh

Project kits, page 136

Safety Matters

- Wear a respirator when sifting enamel.
- Use ventilation when enameling, soldering, and pickling.

What You Do

1. Decide what shape you want to make the earrings. Use the jeweler's saw to cut this shape out of the 22-gauge copper sheet. (You also can use a piece of precut commercial copper.) Slightly form the metal in a die press or with dapping tools.

2. Clean the copper until water "sheets" off its surface. Paint a holding agent on the underside of the copper. Sift a counter enamel between all firings. This prevents firescale from jumping up onto the piece during subsequent firings.)

5. Use a glass bush to remove any finger oils you may have left on the surface of the base-coat enamel. Holding one piece by its edges, brush on the holding agent. Sift a thin coat of enamel in a contrasting color onto the binder, and then allow the piece to completely dry. Scratch a design into the sifted enamel using a scribe, brush, or any other tool that produces the desired line quality. Fire the enamel (see photo). Once it's cool, file or sand the firescale off the edges of the fired piece.

6. Use a glass brush to clean the front of the piece. Apply a holding agent, and then sift on a thin coat of a third enamel color. (You can use a stencil if you wish to cover only a portion of the piece.) Once the holding agent is dry, you can scratch through the most recent sifting to reveal the base coat, second coat,

or both enamel layers. Fire the enamel, let it cool, and then file or sand its edges (see photo).

7. If you want the earrings to have a matte finish as shown, use a commercial glass etching cream. Follow the manufacturer's instructions to apply the cream and process the etching. Wear gloves when working with the etching cream and be sure to wash the finished piece carefully.

8. Repeat steps 1 through 7 to create a second enameled component for the earrings.

9. To make a bezel that fits the enamel, first wrap a piece of string or a strip of masking tape around the piece to use as a guide. Measure the length of the string or tape and add three times the thickness of the bezel wire. Cut a bezel wire to the total measurement. File the ends of the bezel wire perfectly flat. Bring the ends together, making sure they are flush from all angles. Solder the bezel with hard solder as shown, let it cool, and pickle it to remove flux.

10. If the bezel has lost its shape, re-form it around the enameled piece as shown. If the finished bezel is a little too small, enlarge it by placing it on a ring mandrel and gently hammering it with a small steel or brass hammer. If there are big gaps in between the enameled piece and the edge of its bezel, it's too large, You'll need to cut out a bezel piece at the seam and re-solder.

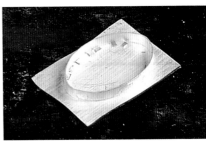

11. Sand the fitted bezel on a flat surface, making sure it's perfectly level on one side. Flux the 24-gauge sterling silver sheet and put the bezel on top. Place small chips of medium solder around the outside of the bezel (see photo). Put the 24-gauge sterling silver sheet on a soldering screen. Heat the piece from underneath to solder the bezel to the sheet without melting the bezel wire. Cool and pickle the bezel. Use the jeweler's saw to remove the excess silver from around the edges of the bezel. Sand the cut edges smooth and flush.

12. Use the rectangular wire to make a decorative ellipse shape to go behind the bezel. Cut it slightly longer than the bezel wire. Bring together the ends of the rectangular wire and solder them closed with hard solder. Let the soldered wire cool, and then pickle it. Remove the excess solder from the seam with a file or sandpaper. Remove any inconsistencies in the soldered wire with a ring mandrel and mallet, and then shape the ellipse to match the silhouette of the bezel.

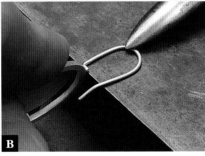

13. Solder the ellipse shape onto the back of the bezel setting with medium solder as shown, and let the piece cool. Pickle it, and then file or sand off any excess solder.

14. Cut a length of 20-gauge sterling silver round wire to use for the ear wire. As shown in photo A, forge one end of the wire to create a flat area for soldering the ear wire to the earring. (The wire end that isn't forged goes through the ear.) Solder the flat wire end of one ear wire to the top of the ellipse shape

with easy solder. Let the settings cool, and then pickle. Sand all of the surfaces to an even finish, and clean the rest of the ear wire with steel wool. Burnish the wire against a steel block to make it stronger. (During soldering, the ear wire was annealed and softened.) Using pliers or your fingers, bend the wire into a hook shape (photo B).

15. Repeat steps 9 through 14 to create a second bezel setting for the earrings.

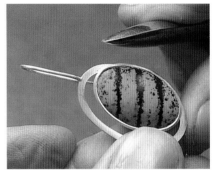

16. Check and correct the bezel heights before setting the enamel pieces. The bezel wire should be just high enough to hold the enamel in place once burnished. The wire should not be so high that it pleats unevenly when burnished. Sand or file the bezels to the correct height against a flat surface as needed. One at a time, place an enameled piece in a bezel. Use a burnisher or bezel pusher to push down the bezel as shown. Press the metal over the edge of the enamel at 12 o'clock; move down to 6 o'clock; across to 9 o'clock; and then over to 3 o'clock. When the enamel is secure and the bezel is pushed down at these four points, smooth it with the burnisher. Finish the earrings as desired with polish, steel wool, or a brass brush.

VARIATIONS

Left: Nick Mowers *Stone and Enameled Earrings No. 2*, 2004. 1½ x ½ in. (3.8 x 1.3 cm). Sterling silver, enamel, copper, 14-karat gold, coral.
Center: Linda Darty and **Corey Fong** *Sgraffito Earrings*, 2003. 1 in. diameter (2.5 cm). Copper, enamel, sterling silver.
Right: Nick Mowers *Stone and Enameled Earrings No. 1*, 2004. 1½ x ½ in. (3.8 x 1.3 cm) Sterling silver, enamel, copper, 14-karat gold, onyx.
Photo by Keith Wright.

Sugar-Fired Necklace

Sharon Massey

To create this necklace you'll enamel copper components and set them in silver bezels.

Techniques

Under-firing sifted enamel for a sugar-fired effect

Bezel-setting enameled copper

Joining several enameled components to form a larger piece

What You Need

Copper sheet, 22 gauge

Fine silver bezel wire, 26 gauge

Sterling silver sheet, 24 gauge

Sterling silver round tubing

Sterling silver round wire, 12 gauge

Necklace clasp, handmade or purchased

Opaque enamels

Sifter, 80 mesh

Project kits, page 136

Safety Matters

• Wear a respirator when sifting enamel.

• Use ventilation when enameling, soldering, and pickling.

What You Do

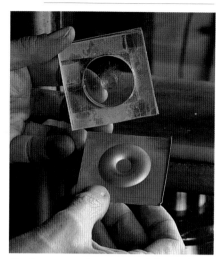

1. Create a dimensional form from the 22-gauge copper sheet. (This project features die-formed cop-

per shapes made with a hydraulic press, and then sawn out.

2. Thoroughly clean the formed copper shape until water sheets off its surface. Paint a holding agent on the back side of the copper form. Sift counter enamel over the binder, let dry, and then fire the enamel until fused. Using a green kitchen scrub pad under running water, lightly scrub the firescale off the front side of the formed copper.

3. Apply a base coat of enamel to the front side of the formed copper piece. (A base coat is necessary for successful sugar firing.) Let the base

coat completely dry, and then fire the enamel until it's fused with a glossy surface.

4. Paint a holding agent on the front side of the piece, and sift an opaque enamel over the fired base coat as shown. (Darker colors are easier to sugar-fire than light ones, and opaque colors usually look better than transparent colors, but you should experiment to find the ones you like.)

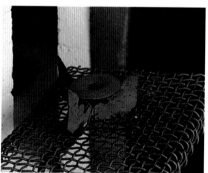

5. When the holding agent is dry, put the piece in the kiln and watch it closely. You'll be removing the piece from the kiln much sooner than you would for a regular firing. Check the progress of the firing very often because the sugar-fire stage occurs only for an instant. Watch the sifted enamel for the slightest change in appearance (this change varies from color to color), and then immediately remove the

VARIATIONS

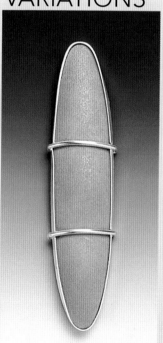

Left: **Sharon Massey** *Brooch*, 2003. 3 x 1¼ in. (7.6 x 3.2 cm). Copper, enamel, sterling silver. Photo by Keith Wright.

Right: **Sharon Massey** *Enamel Brooches*, 2003. 1½ to 2 in. diameter (3.8 to 5 cm). Enamel, copper, sterling silver. Photo by Robert Diamante.

piece from the kiln. Using long steel tweezers, carefully scratch an edge to see if the enamel grains have adhered to the base coat. If the enamel scratches off, put the piece back in the kiln and repeat the firing process. If the enamel stays in place, you've successfully sugar-fired. (If the enamel passes the sugar-fire stage and becomes orange-peel fired, leave the piece in the kiln until the enamel is glossy and try again.)

6. To create a bezel for the sugar-fired enamel, follow step 9 of the Stencil and Sgraffito Earrings on page 138.

7. Following steps 10–11 of the Stencil and Sgraffito Earrings on page 138, flux the 24-gauge sterling silver sheet and solder the bezel to it.

8. Pickle the silver bezel setting, and then rinse it in water and dry

it thoroughly. Use a jeweler's saw to cut off the excess silver from the outside of the bezel (see photo). File and sand the cut edges smooth.

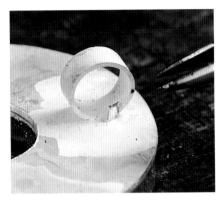

9. Use the jeweler's saw to cut two pieces of the silver tubing. Solder the tubing pieces on opposite sides of the back side of the bezel setting as shown. Pickle the silver setting, rinse, and dry it. Sand the soldered edges to remove any file marks and scratches.

10. Repeat steps 1 through 9 to make as many enameled components and bezel settings for the necklace as you wish. You can make these components different sizes, colors, and shapes as desired.

11. Using the 12-gauge silver wire, make one jump ring for each tubing piece used on your necklace settings. (Jump rings made from such

a heavy wire are so strong that you won't need to solder them closed.)

12. Join the bezel settings together into a necklace by threading one jump ring through each tubing piece on the back side of each setting. Fabricate or buy a clasp to secure the necklace.

13. After linking the entire necklace, set one enameled piece into each bezel setting. Use the method described in step 16 of the Stenciled & Sgraffito Sifted Earrings on page 139 to check the height of each bezel, and file as needed. Following the "clock" method, burnish the bezel settings over the enamels.

Layered Enamel Ring

Kathryn Osgood

Fabricate and enamel a floral centerpiece to crown a sterling silver ring.

Techniques

Sifting and wet-inlaying enamel on a three-dimensional form

Enameling transparent color on fine silver

Using mica as a firing support

Fashioning a simple band ring from wire stock

Cold connecting an enamel using a screw or rivet

What You Need

Fine silver sheet or copper sheet, 22 gauge

Brass screw and nut (090, or size of piece to be attached)

Round sterling silver wire, 10 gauge (length depends on ring size and stem size)

Sterling silver sheet, 22 gauge, 2 x 2 inches (5 x 5 cm)

Bezel cup, 3 mm in diameter, commercial or handmade

Gemstone of your choice, 3 mm

Transparent or opaque enamels

Project kits, page 136

Safety Matters

- Wear a dust mask or respirator when sifting enamels.
- Wash your hands after handling enamels.

What You Do

1. Design the layered metal forms you want to create. (For this ring, the designer formed a three-dimensional flower with three layers of shaped petals. Your metal layers could include flowers, buds, leaves, or even geometric shapes.) Using your sketches as a guide, construct a three-dimensional paper model of the design. Adhere the paper model parts to the 22-gauge fine silver or copper sheet with rubber cement, and let dry. Saw out the shapes with a jeweler's saw.

2. Mark the centerpoint of each metal shape. Use a bit the same diameter as the brass screw to drill a hole through each of the marked points. (Each hole should be tight enough so that the metal pieces thread onto the screw rather than slide onto it.) File all cut metal edges and sand all surfaces and edges with 320-grit sandpaper. Using small round-nose pliers, bend each metal piece to the desired shape (see photo). When bending and forming, make sure the shapes will nest together correctly.

3. Sift a counter enamel onto the back side of each metal form. (Because the pieces are so small, it's easier to place the shapes on a sheet of mica for support rather than trivets.) Fire the metal forms until the enamel is fused, and let them cool.

4. Touching only the edges, turn over the metal pieces. (If you're working on fine silver and not copper, no firescale will form on the metal edges, so if you don't touch pieces, no cleaning or filing

between firings will be necessary. If you're working with copper, clean off the firescale prior to applying the base coat.) Apply clear enamel or any base coat color you desire to the front side of each metal layer as shown, and fire the pieces.

5. Continue to apply transparent or opaque enamels until you achieve the colors you desire. If you prefer, you can build up color by applying wet enamels with a brush.

A

6. Use the 10-gauge round sterling silver wire to make a simple band ring. Cut the wire to the desired size and bend the ends until they meet. Flux the wire ends, solder

VARIATIONS

Above: Kathryn Osgood *Set of Four Flower Rings*, 2004. 1½ x 1 x 1 in. (3.8 x 2.5 x 2.5 cm). Copper, enamel, sterling silver. Photo by Linda Darty.

Right: Kathryn Osgood *Mourning Locket* (detail), 2003. 1⅜ x ⅝ in. (3.5 x 1.6 cm). Copper, silver, enamel, gemstone. Photo by Robert Diamante.

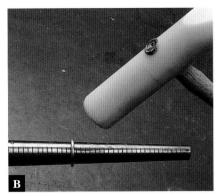

them together with hard solder (photo A), and then pickle the band. Using a ring mandrel and a rawhide mallet, hammer the ring until it's round (photo B).

7. To form the stem of the ring, cut a piece of the 10-gauge sterling silver wire. File one end of the cut wire to match the curve of the wire band, so the stem fits snugly against the body of the ring. Using a third arm to support the ring and the stem, flux the joint and solder with medium solder. Let the band cool, and pickle it. File and sand the ring to clean the soldered seam.

8. File the octagonal nut that fits over the screw until it's round and the desired size. Cut a strip of the 22-gauge sterling silver that's long and deep enough to enclose the nut. Form and solder this strip into a bezel. Sand the bottom of the bezel smooth. Using hard solder, solder the bezel to a piece of 22-gauge sterling silver sheet that is slightly larger than the bezel's diameter.

9. File and sand the bezel edges smooth and round. File the top of the stem level. Flux and solder the bezel to the top of the stem with medium solder as shown.

10. Use easy solder to solder the rounded nut into the inside of the bezel as shown. Pickle the ring, and then use a brass brush and liquid soap to achieve a satin finish on the ring.

11. Use a commercial 3-mm bezel cup or construct one to fit your stone. Use easy solder to solder the bezel cup to the top of the brass screw. Set the stone following the method described on page 139, step 16.

12. To determine the correct length to cut the screw, pre-assemble the enameled piece. Thread each layer onto the screw so the stone center fits snugly against the top hole as shown. Add the measurement of the depth of the nut to this length. Remove the enameled pieces from the screw. Use the jeweler's saw to cut the screw to the correct length.

13. To assemble the ring, thread the enameled metal onto the screw until the bezeled stone is snug against the top layer. Thread the screw into the nut (see photo). Threading a tiny rubber washer onto the screw before joining will give the finished ring a tighter fit.

Etched & Enameled Brooch

Linda Darty

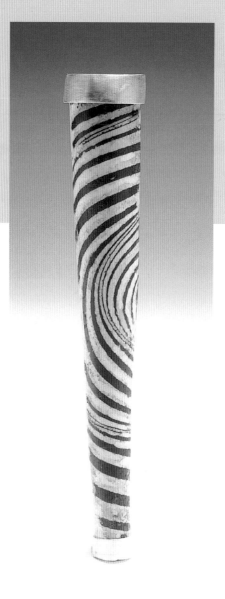

This unique brooch features an excellent transfer paper etching resist.

Techniques

Using circuit-board heat-transfer paper

Etching copper in an acid bath

Enameling with liquid enamel

Making cap settings

Using an colored-epoxy adhesive to match enamel

What You Need

Copper sheet, 18 gauge

Fine silver sheet, 24-gauge

Stainless steel or 14-karat gold pin stem wire

Tubing to fit pin stem wire

Black and white image

Circuit-board heat-transfer paper

Ferric chloride acid

Ammonia

Powdered and/or liquid enamels

Glass etching cream (optional)

Epoxy resin, two-part, long-drying

Paint pigment to match back side of brooch

Wax paper

Photocopier or laser printer

Wood block

Household iron

Project kits, page 136

Safety Matters

- Wear a mask when sifting
- Work with acids in a well-ventilated area
- Wear gloves and eye protection when etching

What You Do

1. Photocopy or laser-print a high-contrast black-and-white image onto the dull side of the circuit-board heat-transfer paper as shown. (You may want to first make a test on plain white paper. If so, experiment with the settings on the copier or printer to darken the image and to heighten the contrast without creating gray shadows in the white areas.) If the image is small, make several copies of it on plain white paper, and then cut and paste them together to fill up an entire transfer sheet. In case you have difficulty transferring the image to the copper the first time, it's good to have extras with which to practice.

2. Design a brooch shape on paper, and then use it as a pattern for the metal. Cut a very flat piece of 18-gauge copper that is larger than the size of the brooch design. (The copper must be very flat so you can easily iron it. I prefer to cut off the scrap metal after etching the design. If the copper is cut to the proper shape prior to etching, its edges would need a resist or they would be "bitten" by the acid.) Thoroughly clean the metal under running water with sandpaper, a green scrub pad, or pumice, until water sheets across the surface.

VARIATIONS

Linda Darty *Three Etched & Enameled Brooches*, 2003. 3 x ¾ in. (7.6 x 1.9 cm) each. Copper, sterling silver, enamel. Photo by Keith Wright.

3. Position the image side of the circuit-board heat-transfer paper face-down on the metal. Place the metal on a wood block and use an iron set on or near a high setting to transfer the image (see photo). Hold the iron on the paper for a few seconds, making sure it starts to tack down before moving the iron to burnish the rest if the paper. If the iron is too hot the paper will bubble up as the plastic starts to melt. If this happens, turn the iron down and keep burnishing. If you iron the paper too long with too much heat, the black image will bleed on the metal surface, and the lines won't be crisp. If the transfer doesn't work, begin again with a new piece of paper and try using a cooler iron. If you continue to have difficulty transferring the image to the metal, there may be a problem with the toner in the photocopier or printer you're using, so try a different machine.

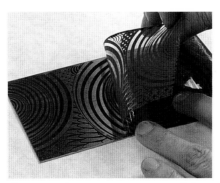

4. Once the image is successfully transferred and the metal is completely cool, peel off the transfer paper (see photo). If you notice any residue from the transfer paper left on the metal, roll up a piece of masking tape and gently press it on the surface to lift the residue. You can use a sharp craft knife or razor blade to scrape away unwanted black areas and clean up the lines. If there are areas that did not transfer, use an oil-based paint pen to touch up the resist.

5. Etch the metal in ferric chloride following the method described on pages 114–119. After etching, immediately put the metal in a strong solution of ammonia and water to neutralize the acid residue. Scrub the metal with a toothbrush until the mustard-colored residue disappears. (If you heated the metal with this ferric chloride residue left on its surface, you would create toxic fumes.) Use acetone or sandpaper to remove the transfer paper from the clean metal surface as shown.

6. Rubber-cement the paper brooch pattern to the etched metal or simply trace the pattern with a permanent fine-tip marker. Cut the etched copper to the desired brooch shape.

7. Use a mallet to form the metal to the desired shape. You could make a spiculum or use a hydraulic die press or dapping blocks to form the metal.

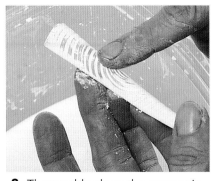

8. Thoroughly clean the copper in preparation for enameling. If the etching is shallow and you want to create a basse taille brooch, simply sift on transparent color. (See the project variations for an example of this approach.) To create a brooch similar to the project, dip the piece in white liquid enamel. When dry, brush powdered enamel off the high spots with your fingers (see photo). Fire.

9. Examine the fired piece and decide if you're satisfied with the enameling. If desired, you can sift

and fire transparent color over the piece so it will be smooth, and you'll see the pattern from the enamel and the copper beneath the glass. (I sifted clear enamel over the entire piece, so the firescaled copper areas would turn reddish brown.)

10. To create a matte surface, paint glass etching cream on the enamel and let it sit approximately five minutes.

11. Cut a strip of the 24-gauge fine silver sheet to make a sleeve for the top of the brooch. Shape the strip to tightly fit the brooch. Solder together the ends of the strip with hard solder to make an end cap. Repeat this step to make an end cap for the bottom of the brooch.

12. Using medium solder, solder each end cap to a piece of 22-gauge silver sheet. Trim away excess metal, and then file and sand the cut-and-soldered edges.

13. Construct brooch findings to hold the pin mechanisms. Solder one finding onto the back side of the top and bottom end caps.

14. Check to see that the end caps still fit the brooch. If you want to solder decorative tops on the caps, do so now. (For example, you could attach bezels for gemstones or decorative metalwork.)

15. If the fit is good and all the parts are cleaned and finished the way you want them to be, mix up two-part long-drying epoxy resin on a sheet of wax paper. Add dry paint pigment to the epoxy resin as shown so it matches the enamel color on the back of the brooch.

16. Tilt the assembled brooch so the resin won't flow out, and then fill one end cap with the epoxy resin (see photo). Let the brooch sit for several hours or overnight. Turn the brooch over and fill the other end cap with the resin. Let it completely dry.

17. If the shape of your brooch is conical (similar to the featured project), you can burnish down the end caps to further secure them to the brooch.

18. Insert the pin stem wire into the top finding and bend it to the correct shape. Cut the wire to the correct size, and then use a file and carborundum paper to sharpen the end to a point.

Painted Buttons

Mi Sook Hur

Make very painterly buttons using water-based acrylic and watercolor enamels.

Techniques

Painting fine details with acrylic and watercolor enamels

Setting an enamelled piece in a hollow formed frame setting

What You Need

Copper sheet, 18 gauge

Sterling silver sheet, 20 gauge

Sterling silver wire, 12 gauge

Acrylic enamels

White enamel for base coat, 80 mesh

Sable paintbrushes, in assorted sizes such as 000, 00, and 0

Project kits, page 136

Safety Matters

• Work in a well-ventilated area

• Wear a dust mask or respirator

What You Do

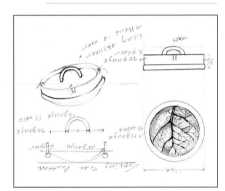

1. Make quick thumbnail sketches of button design ideas. Use watercolors and colored pencils on paper to develop a drawing and give you an idea of how you want the finished enamel button to look.

2. Use a jeweler's saw to cut out a copper disk, and then file its edges smooth. (You also could purchase a pre-cut piece of copper.) Place the copper disk inside the convex dome of the dapping block. Place the dapping punch over the copper and sharply hit it with a hammer or mallet to form the copper.

3. Thoroughly clean the domed copper disk. Apply a counter enamel to the back side of the disk, let dry, and fire. Clean the firescale off the front side of the disk. Apply the white enamel base coat, let dry, and then fire. Once the disk is cool, file its edges and clean it well with a glass brush under running water.

4. If you want to paint the design onto the button freehand, skip to step 5. Otherwise, place red dressmaker's paper on the disk with its waxy side facing the white enamel base coat. Place the drawing over the transfer paper. Using a sharp pencil, trace the drawing to transfer it onto the coated metal. (Any color of transfer paper can be used. The red just shows up well on white enamel.)

5. Use acrylic enamel to paint the outlines of the drawing as shown. (In this case, light brown mixed with a little water.) Keep the lines as thin as possible. They can be made darker with enamel coats applied later . Let the painting completely dry. Fire the disk in the kiln until the surface becomes shiny. (It will look matte before firing.)

6. Repeat step 5 to create additional layers of color. It's best to start with light color enamels, adding them in thin layers, and working toward dark color enamels as shown in photo A. (The artist made light green enamel paint by mixing green with yellow and adding a bit of water. Adding more water gives the paint more transparency. She left some of the white base coat visible to serve as highlights.) Continue painting, adding the medium values next, and then the darkest values. Always apply the acrylic enamel thinly and fire the disk between each coat.

7. Following the method described in step 5, paint in the details of the drawing. (A small, fine round brush was used to draw the tree branches and trunk. The artist used two tones of brown enamel, formed by mixing green, brown, and black together. For painting the trunk, she used the darkest brown. For painting the branches, she used a lighter tone made by mixing more green enamel and water into the dark brown.) When the enamelled disk is painted and fired to your satisfaction, file its edges smooth.

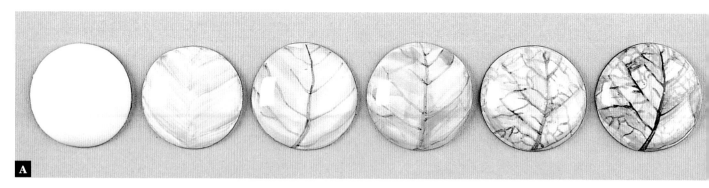

A

VARIATIONS

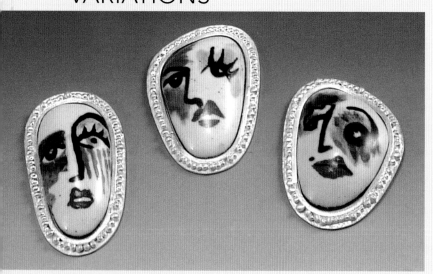

Above: Mi Sook Hur *Painted Buttons*, 2003. 1¾ in. diameter (4.4 cm) each. Copper, enamel, sterling silver. Photo by Keith Wright.

Left: Lindsey Hardin *Face Buttons*, 2003. 1 x ¾ x ½ in. (2.5 x 1.9 x 1.3 cm) each. Silver, enamel. Photo by Linda Darty.

8. Cut a strip of the 20-gauge sterling silver that is the approximate height of the enamelled disk. Form the strip to the disk's outside diameter. Flux and solder together the ends of the formed silver strip with hard solder.

9. Form a piece of the 12-gauge sterling silver round wire to fit the inside diameter of the soldered silver strip. Solder the wire to the strip as shown.

10. Place the enamelled disk into the silver setting and check its fit (see photo). File the edges of the disk as needed to form a good tight fit.

11. Use a scribe to trace the shape of soldered silver setting onto a sheet of 20-gauge sterling silver. Use a jeweller's saw to cut out the traced disk. Use needle-nose pliers to create a half-circle jump ring

from 12-gauge sterling silver wire. Flux and solder the half-circle jump ring to the back of the silver disk so the button can be stitched to fabric.

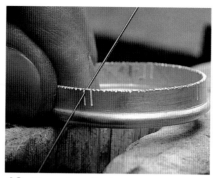

12. Use a jeweller's saw to cut tiny tabs in the top edge of the silver frame setting.

13. Place the enamelled disk into the frame setting, and then place the silver disk behind it (photo B). Use a burnisher to bend the tabs over the disk to hold the piece together (photo C). (You may need to file the tabs thinner before pushing them over.)

Liquid Enamel Pendant

Robert W. Ebendorf

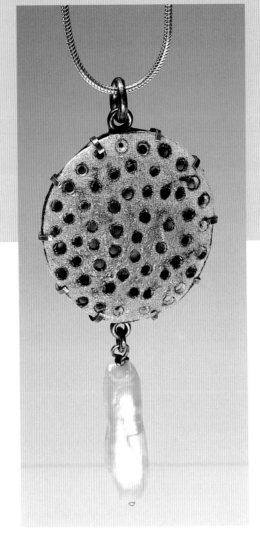

Fashion a simple pendant, and then apply liquid enamels and sifted transparent colors to make it a work of art.

Techniques

Applying liquid enamel in combination with sifting techniques

Experimenting with controlled over-firing

Tab-setting an enameled form

Making a simple bail for a pendant

What You Need

Copper sheet, 18, 20, or 22 gauge

Wire, 12 gauge (for jump ring)

Gemstone or pearl

Handmade or found chain

Transparent enamel

Water-base liquid enamel, white or transparent

Project kits, page 136

Safety Matters

• Wear a dust mask or respirator when sifting enamels

• Wash hands well after handling enamels

• Solder in a well-ventilated area

What You Do

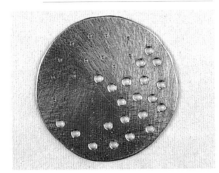

1. Determine what shape pendant you want to make. (Cutting out a paper pattern may help you refine your design. You can then adhere the paper pattern to the copper sheet with rubber cement, making it easy to follow the pattern while sawing around its edges.) Saw out

the copper pendant shape. Using a hammer and a sharp steel nail or a centerpunch, dimple the metal and then drill decorative holes to create a pleasing design. (You could use other techniques to pattern or texture the copper.)

2. Form the metal (see photo). This piece was formed in a wooden dapping block with a hammer and a steel dapping tool. Prepare the copper for enameling by gently heating it to burn off oxides. Pickle the copper to remove firescale from the surface. Rinse and dry the metal.

3. Prepare the liquid enamel to a consistency that is a slightly thicker than heavy whipping cream. Test the liquid enamel by brushing it on the copper. If the enamel is too thin, it won't coat the copper well. If it's too thick, it won't flow into the low areas of the surface design. Use any size brush to paint liquid enamel on both sides of the copper pendant, or dip it into the liquid. Place the enamel-coated metal on a trivet to dry.

4. If you wish to incorporate firescale into the design, you need to make the liquid enamel thinner in areas. Simply use your finger to brush off the enamel. When over-fired, a green copper oxide color will appear where the enamel is thinner.

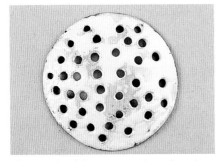

5. Use the kiln or a torch to fire the enamel until you achieve the color and surface you desire (see photo). File the edges of the pendant, and then clean its surface with a glass brush under running water.

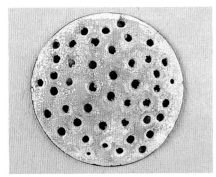

6. Fire as many thin enamel layers on the pendant as you want. Select pleasing transparent colors and sift them over the fired liquid enamel in interesting areas as shown. (In this piece, transparent beige was sifted on and fired until fused. The last layer of enamel was under-fired to leave an orange peel surface.) After firing, clean and finish the pendant edges.

7. Lay the pendant on a sheet of paper and trace its contour. Transfer this shape onto a copper sheet. Using dividers with a sharp scribe, mark even sections around the perimeter of the piece so that tabs can be cut out to create the setting.

8. Solder jump rings at the top and bottom of the setting. Drill holes through the middle of the jump rings to create a hanging mechanism for the chain and a ring to

hang something from the bottom of the pendant. Saw around the jump rings and all the tabs (see photo).

9. Carefully sand and finish the metal before pushing the tabs over the enamel to set the piece as shown. Hang a gemstone or a pearl from the ring in the bottom of the setting. Place the enameled pendant on a commercial or hand-made chain.

VARIATIONS

Melissa Manley (left) and **Robert Ebendorf** (right) *Liquid Enamel Pendants,* 2003. 1½ in. diameter each (3.8 cm). Copper, liquid enamel, pearls, found chains. Photos by Keith Wright.

Fold-Formed Vessel

Tim Lazure

Create a distinctive bowl by fold-forming copper and covering it with liquid enamel.

Techniques

Using fold-forming techniques to create textured surfaces

Sinking a metal bowl

Dipping metal into liquid enamel

Over-firing on a fold-formed surface

What You Need

Copper sheet, 22 to 14 gauge

Liquid enamels for copper

Large bench vise with jaws at least ½ the diameter of the bowl you wish to make

Deadblow, plastic, or wooden mallet

Cross-peen hammer or chasing tools and a ball-peen hammer

Large mushroom stake

Plastic bowl, bucket, or container to hold liquid enamel (must easily accommodate size of copper bowl)

Project kits, page 136

Safety Matters

• Work in a well-ventilated area

• Wear appropriate shoes or boots when using hammers and stakes to protect your feet in case they are dropped

• Wear a dust mask or respirator

What You Do

1. Design a basic contour for the bowl you want to make. (Although this bowl was made from a circular disk, you could use an oval copper piece, something more asymmetric, or even a square or a rectangle.) To determine what size copper disk you'll need, add the average width of the bowl to its height. This figure is the diameter of the disk. Set the spacing of the dividers or compass to the disk's radius (half of its diameter). Scribe or draw the circle onto the copper sheet. (A scribed line is a little more precise, and it won't rub off when you saw out the disk.)

2. Use a saw frame with a deep throat to cut out the copper shape marked in step 1. Saw on the outside of the line to leave room for filing. Slightly file the cut edges to remove all burrs.

3. Fold-forming gives sheet metal volume, shape, and texture. To make the first fold easier, anneal the copper disk by heating it with a torch until it glows cherry red, and then quench in water.

4. Once the copper is cool, use your hands to fold it approximately in half, leaving a soft bend in the metal.

5. With its folded side facing up, place the copper between the jaws of the vise. Angle the metal so the fold is about 1 inch (2.5 cm) above the vise jaws on one side and about

¼ inch (6 mm) above the jaws on the other side as shown. Tighten the jaws of the vise to bring the metal together while leaving the bend with a loose radius. This produces a loosely folded tapered effect. Leave the copper in the tightened vise.

6. Using a deadblow, plastic, or wooden mallet, hammer down onto the loose fold (see photo). Start striking the folds on the ends and slowly work your way toward the middle. Using this method produces a tapered, flattened area with two crisply folded edges. Leave the copper in the vise.

7. Use a cross-peen hammer or chasing tools with a chasing hammer to texture the flattened area created in step 6 (see photo). At this point, a cross-section of the bowl should look like the letter T.

8. Remove the copper from the vise. Turn the metal 90 degrees and

place it back into the vise as if you were closing the vise on the "wing" of a copper "airplane." Use the cross-peen hammer to strike down onto the folded metal edge, leaving a textured edge (see photo). Take the metal out of the vise, flip it 180 degrees, and strike the other fold with the cross-peen hammer. At this point, the heavily hammered metal has become work-hardened. Re-anneal it following the method described in step 3.

9. Use your hands or a mallet and anvil to open the folded copper as shown. (You may wish to use a tool such as a butter knife to help start

VARIATIONS

Above: Kate Cathey *Fold Formed Baskets*, 2003. 4 x 2 in. (10.2 x 5 cm) approx. Copper, liquid enamel. Photo by Linda Darty.

Left: Annie Grimes *Abalone Bowls*, 2003. 4 x 4 x 2 in. (10.2 x 10.2 x 5 cm) each. Copper, liquid enamel. Photo by Linda Darty.

opening the fold. Thinner gauge metal is easier to open. For thicker gauges use a heavier mallet.) Use a hammer to flatten the opened copper on the anvil, flipping it back and forth until it's relatively flat. If you find you're losing too much textural detail, place a piece of scrap leather between the copper and the anvil.

10. When you look at the metal you'll see a heavily textured V-shaped area near the center of the disk. To the left and the right there are two smooth areas. Pick one of the smooth areas and make a crisper fold at any angle you feel is visually interesting. To make this fold, put one end of the copper in the vise and use the mallet to form a crisp 90-degree angle. Once the bend is started in the vise, use the mallet and anvil to finish closing it tightly. If you want to create texture on this folded edge, place the copper back into the vise with the folded side up, and strike the fold with a cross-peen hammer. Continue folding, annealing, and unfolding the metal until you achieve a pattern you like (see photo). Use an anvil with a leather mallet to flatten the metal, and then re-anneal it.

11. Hold the copper at a 30-degree angle over a depression in a stump or wooden block. Make sure to support the metal with your hand and with the stump or block. Holding the tang of the mushroom stake in your hand, strike the metal near its outside edge (see photo). Strike the metal in concentric circles, turning the piece at an even pace. (This piece was hammered on its outside edge about four or five times. This number may vary depending on your strength and personal technique.) Slowly move toward the center of the copper, keeping the bowl's contour as even as possible. (Use a wooden or plastic mallet if you prefer.) The bowl will work-harden as you hammer. Make sure to re-anneal the metal while forming it into the shape you desire.

12. Secure a mushroom stake in the vise. (The closer the stake fits the interior contour of the bowl the

better.) Place the bowl on the stake and strike it with a leather mallet as shown until the bowl is round and lump-free. (This technique is called *bouging*.) Be careful not to bouge too much or you might loose some of the bowl's textural detail.

13. To clean the finished bowl, anneal, pickle, and rinse it until water sheets off its surface. (If you want to keep some firescale on the piece as a design element, leave the bowl in the pickle briefly.)

14. Hold the bowl by its edges and dip it into the liquid enamel, shaking off any excess. Place it upside-down on a large trivet. As the enamel dries you may need to touch up places that were marked by your fingers. Dab a little more enamel into those areas with a paintbrush. Make sure the enamel has dried completely before you place it into the kiln. Once the enamel is dry you can rub some areas thinner or smooth drips with your fingers.

15. Fire the enamel. (Because of all of the texture, ridges, and valleys on this bowl, the designer preferred to over-fire his project. This resulted in more color variation and emphasized the recesses in the folded surface. This bowl was fired three times.)

16. Finish the edges of the bowl as you wish. You can file them, sand them, or leave them coated with liquid enamel.

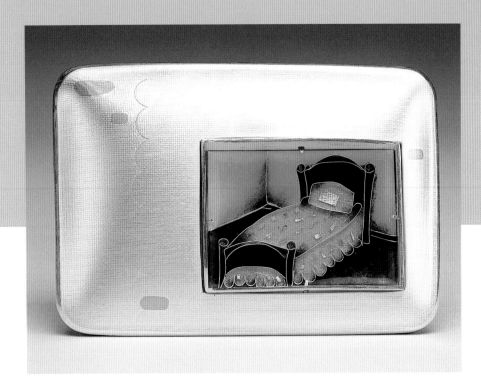

Cloisonné Brooch

Adrienne Grafton

Tell your own story with a narrative cloisonné enamel.

Techniques

Enameling on fine silver with transparent colors

Using cloisonné wires to create a narrative image

Creating a die formed-and-tabbed brooch setting

What You Need

Fine silver, 20 gauge (size depends on your design)

Cloisonné wire, fine silver

Fine silver foil or gold foil

Sterling silver, 20 gauge (size depends on your design)

Brooch findings, handmade or commercial

Double-sided transparent tape

Transparent enamels (washed and tested on fine silver)

Clear thermoplastic, 5 x 5 x ¼ inch (12.7 x 12.7 x .6 cm)

Matte board or glass with polished edges, 4 x 4 inches (10.2 x 10.2 cm)

Spiral saw blades, for cutting thermoplastic

Wax files, for filing thermoplastic

Hydraulic die press

Project kits, page 136

Safety Matters

- Work in a well-ventilated area

What You Do

1. Draw a simple outline for the enameled element. (Here a rectangular shape was used.) Design a cloisonné wire structure to fit inside the outline. Experiment with colored pencils to determine what enamel colors you want to use in the piece. Copy this design onto tracing paper, and tape the paper onto the 4 x 4-inch (10.2 x 10.2 cm) piece of matte board or glass. Put a layer of double-sided tape over the drawing so the cloisonné wires will stick to the tape when you later bend and place them.

2. Design a setting that relates to the image you created with cloisonné wires in step 1. (The artist used a pillow shape to relate to the bed created with cloisonné wires). Keep in mind that the shape for the setting will be die-formed.

3. Trace the exterior contour of the setting design onto the 20-gauge fine silver sheet and saw it out. File the cut edges smooth with needle files. Clean the metal to prepare it for enameling, and then handle it only by the edges. Wearing a dust mask, sift a base coat and counter enamel of clean clear enamel onto the fine silver.

4. Using the matte board or glass template for reference, begin bending the wires to follow the lines on the tape. Use your fingers, tweezers, and pliers to help form the cloisonné wire and use sharp scissors to cut it.

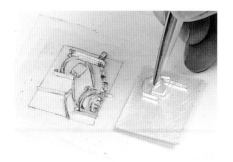

5. Use sharp, clean tweezers to dip the cloisonné wire pieces into the holding agent and transfer them to the front side of the enameled fine silver. Let the holding agent dry.

6. Fire down the cloisonné wires and check them to be sure they adhered to the base coat as shown. (Refer to page 112, step 4 for a more detailed description of this process.)

7. Prepare a palette of washed, wet transparent enamels. Inlay a thin coat of enamel into the cloisonné cells as shown. Let the enamel dry, and then fire the piece to the orange peel stage. (This under-firing prevents the wires from sinking.)

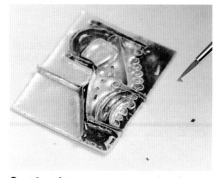

8. After firing one or two thin layers of colored enamel, apply fine silver foils to the piece as shown. (Refer to pages 80–84 for more information about working with foils.) Allow the piece to dry, and then fire.

9. Continue to apply and fire thin layers of transparent enamel until the glass reaches the top surface of the cloisonné wire. Once it's flush, abrade the surface with a diamond stick under running water until it's even. Use 400-grit wet emery paper to polish the wires, and then a glass brush to clean them.

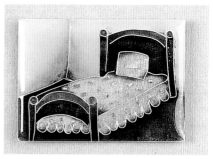

10. Fire the enameled piece one last time to gloss its surface (see photo). Trace the contour of the enameled piece onto the clear thermoplastic to make the die for the setting.

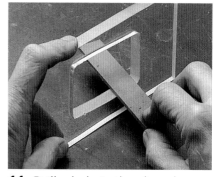

11. Drill a hole in the clear thermoplastic inside the traced shape. Use a spiral saw blade to cut out the shape. File the cut edges smooth (see photo).

12. Texture or cleanly sand the 20-gauge sterling silver you plan to die-form for the setting. Anneal and pickle the metal, and then form in a hydraulic die press.

13. Cut a strip of 20-gauge or 22-gauge sterling silver that is exactly long enough to fit around the cloisonné piece and wide enough to be higher than the depth of the die-formed silver. Carefully drill small holes in at least four locations on this strip. Thread the saw blade through each hole and saw lines. These holes and sawed lines form the tabs that hold the cloisonné enamel in place.

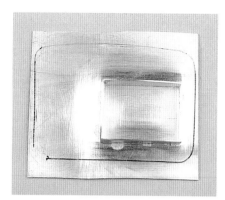

14. Form the silver strip into a frame that fits the cloisonné enamel. Solder the frame with hard solder. Use another piece of 20-gauge sterling silver, larger than the one that was die-formed, to make the back side of the brooch. Solder the frame to this backing with hard solder (see photo).

15. Cut a hole in the die-formed silver shape that fits perfectly over the frame that is soldered to the backing. File and sand the hole, and then slip it over the frame. Shorten the walls of the frame to the desired height using files and sandpaper.

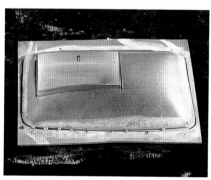

16. Solder the die-formed piece to the backing with medium solder (see photo). Pickle and rinse the piece. Saw, file, and sand its edges.

17. Solder the brooch findings to the back side of the piece and finish the metal as you wish. (A glass brush was used to polish the silver on this piece.)

18. Slip the cloisonné enamel into the frame. Using a slender burnisher, pull the tabs out from the frame onto the cloisonné enamel to hold it in place.

VARIATIONS

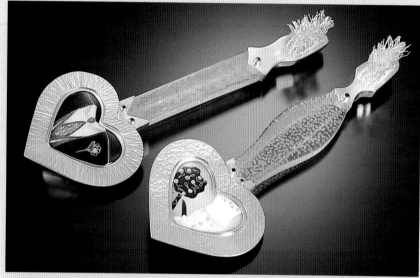

Adrienne Grafton *Just Married Toothbrush Set*, 2003. 7½ x 2½ x 1 in. (19 x 6.4 x 2.5 cm) each. Fine silver, sterling silver, enamel, garnet; cloisonné. Photo by Robert Diamante.

Bryan Park *Prayer Ring*, 2002. 3 x ¾ x 2 in. (7.6 x 1.9 x 5 cm). Sterling silver, copper, sodalite, jade, pearls; cloisonné. Photo by Keith Wright.

Melissa Manley *The Death of Prince Charming*, 2003. 6 x 6 in. (15.2 x 15.2 cm). Sterling silver, copper, dried toad, mica, brass, tulle; cloisonné. Photo by Robert Diamante.

Etched Champlevé Teaspoon

Kate Cathey

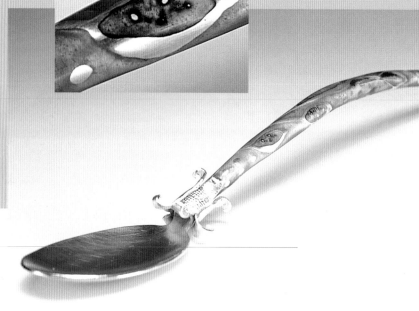

Use etched champlevé and basse taille enameling techniques to create this lovely silver teaspoon.

Techniques

Enameling on fine silver with transparent colors

Acid-etching metal for champlevé

Kiln-soldering metal connections

What You Need

Fine silver sheet, 18 gauge

Sterling silver sheet, 14 gauge

Eutectic solder

Oil-based paint pen or oil-based model paint

Clear plastic packing tape

Transparent enamels for silver

Baking soda

Glass etching paste or liquid solution

Ferric nitrate solution

Magnetic chemical spinner or some other device to agitate acid (optional, see page 116 for suggestions)

Small glass or plastic container to hold acid

Pointed tool to check etching

Project kits, page 136

Safety Matters

- Use an acid-gas respirator, acid-resistant gloves, and eye protection when etching with ferric nitrate
- Work in a well-ventilated area

What You Do

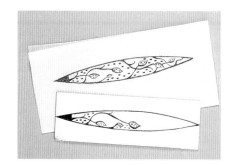

1. Plan the design you want to etch. (This project was created with a spiculum-formed handle though any type of handle will work using similar techniques.) Leave a solid silver border around the edges of the metal to create a frame that will enclose the enamel. Make sure at least ⅛ inch (3 mm) at one end of the handle won't be etched so,

it can be more easily soldered to the piece that attaches to the spoon bowl. Keep in mind that the lines that you draw with the resist on the sheet silver will be the lines that remain raised silver around the enamel on the finished piece. Design the pattern to be etched into the handle on paper, and then transfer it onto the 18-gauge fine silver sheet with carbon paper (see photo).

2. Apply the acid-resistant oil-based paint pen or model paint over the transferred drawing. After about 30 minutes, when the first coat of painted resist is dry, you may want to reapply some areas of the paint. Let the piece dry for two to four hours or overnight.

3. Apply clear packing tape to the back side of the metal to act as an acid resist and to function as a hanging mechanism in the acid bath. Set up the acid bath and etch the piece, referring to the detailed instructions on pages 118 and 119 as needed.

4. Approximately every 20 minutes check the depth of the etching by lightly prodding the resist edge with a pointed tool. Take care not to scratch off any of the painted resist. Remove the metal from the acid bath when the etching is complete. If the resist starts to lift off, dry the piece with a hair dryer and reapply the paint where needed. Rinse the acid etched piece. Scrub it with a paste made of baking soda and water to neutralize the acid. Use sandpaper or a solvent to remove the rest of the resist from the etched metal.

5. Make a paper pattern for the handle shape. Place it on the metal and saw out the handle shape. File the cut edges clean and flat so they will fit together for soldering.

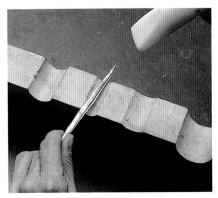

6. Use a hammer to form the metal, annealing it as needed as the metal work hardens. Work slowly and carefully, forming the metal until the edges meet tightly.

7. Solder the handle seam with eutectic solder. If you plan to enamel transparent color over the seam, depletion-gild the solder joint until the black copper oxide from the solder disappears. If you

won't be enameling over the joint, simply file and sand the area clean.

8. Clean the metal well. Use the wet inlay technique to apply the enamel in thin layers as shown.

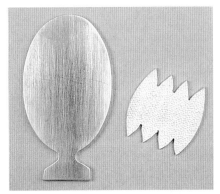

9. Draw the shape of a spoon bowl (left) and a transition piece (right)

on tracing paper. Make sure to incorporate a flange that will be formed to fit inside the transition piece that connects to the spoon handle. Adhere the drawings to the 14-gauge sterling silver sheet with rubber cement and carefully saw out the shapes (see photo).

10. File and form the spoon bowl on a teaspoon stake, on an oval or round mushroom stake, or in a hydraulic die press.

11. Use a small round mandrel with a rawhide or plastic mallet to hammer the flange of the spoon round so it fits inside the transition piece. (The open flange remains unsoldered. This allows you to make

VARIATION

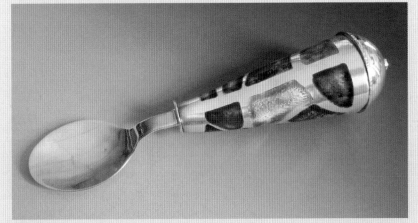

Nate Sealy *Champlevé Spoon*, 2003. Fine silver, Sterling silver, enamel, gemstone. Photo by Linda Darty.

minor adjustments to tightly fit it into the transition piece you'll later form.) File and sand the fully formed spoon bowl.

12. Depletion-gild the silver spoon bowl so it won't produce firescale in the kiln when later soldered to the handle. (It would be difficult to sand and polish the bowl once the handle is attached.)

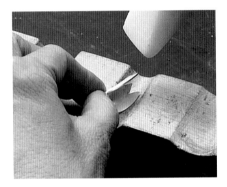

13. The transition piece that connects the handle to the spoon bowl is a soldered ring that fits around the formed flange on the spoon bowl and the un-enameled end of the handle (see photo). This transition piece tightly holds together these two elements so they can be kiln-soldered. You may want to fabricate a plain ring, or you can chase a design, cut the metal edge, or texture the metal in some other way. Make sure the ring fits snugly.

14. Flood the inside of the ring with medium solder. You may have to file out a bit of the solder if the fit to the spoon bowl and the handle is too tight.

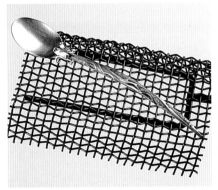

15. Carefully paint the two metal surfaces that need to be soldered with flux. Fit the handle and the spoon bowl together with the ring transition piece. Set the spoon up to be fired on a sheet of mica on top of a firing rack, or directly on a firing rack if the spoon design permits (see photo). Make sure the position is stable, and if your formed handle has a slight curve, that only the end of the enameled handle rests on the mica or rack. Fire the spoon, removing it from the kiln just as you would if you were applying a coat of enamel. Since the enamel surface was sanded to make it level, it will be matte. You'll know you've fired the piece to the correct temperature when the enamel glosses.

16. Decide on a finish for the enamel and the metal. If the glass is to be matte, you can use glass etching cream or solution, or you can sand the surface with fine-grit emery paper. Finish the silver surfaces with steel wool, a soft brass brush, a glass brush, or a rouge cloth.

Champlevé Brooch

Linda Darty

Use saw and solder champlevé with basse taille enameling to create a landscape brooch.

Techniques

Enameling on sterling silver with transparent colors

Using the saw and solder champlevé enameling technique

Working with basse taille texturing

Soldering findings on a finished enamel

What You Need

Sterling silver sheet, 22 gauge, 1 to 2 inches (2.5 to 5 cm) square

Sterling silver sheet, 18 gauge, 1 to 2 inches (2.5 to 5 cm) square

Pin-back findings, commercial or hand-made

Transparent enamels tested on sterling silver

Project kits, page 136

Safety Matters

• Ventilation is necessary when using fluoride-based soldering fluxes, which work better for this technique. Fluoride-based fluxes inhibit firescale at a higher temperature than fluoride-free fluxes.

• Work in a well-ventilated annealing area. When depletion-gilding, the hot pickle solution is caustic and an irritant to the respiratory system.

What You Do

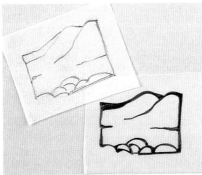

1. Design a line drawing of the saw and solder champlevé piece you want to create. Use tracing paper and a fine-tip permanent marker to redraw the design. To create depth and interest, make the design lines smoothly transition from thick to thin in different areas as shown. (Remember: it's the line you're working with at this point, not the spaces that will eventually hold the enamel. Paying careful attention to these lines, which will be silver in the final design, is important to the success of the piece.) Leave a perimeter of silver around the entire outside edge of the design to function as a built-in setting.

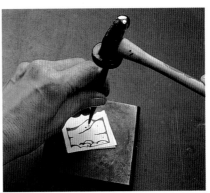

2. Transfer the finished drawing to the 22-gauge sterling silver piece. (I like to use rubber cement and tracing paper. I apply the rubber cement to both the paper and the metal and let it dry until tacky before joining the surfaces.) Center punch (see photo), and then drill holes in all the spaces so you can insert the saw blade through each hole and saw around the drawn lines.

3. When sawing out the spaces, keep the saw blade vertical so the cut metal edges are very straight. You don't want to have to file the pierced edges much. Excess filing could round the edges and cause cracking in the enamel if the wall presses on the glass at an angle.

4. If it's absolutely necessary, carefully file the sawed lines to clean them up. Be sure to keep the file perpendicular to the silver and do not round or angle the edge.

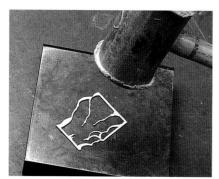

5. Use a leather mallet to hammer the pierced silver piece on a rigid flat surface, such as steel or hard wood.

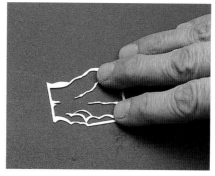

6. Tape a piece of 320- or 280-grit emery paper to a very flat surface.

Rub the pierced silver piece on the sandpaper as shown until it's completely flat and clean on both sides. Pay careful attention to the sanding marks so you can tell if they're even and the piece is truly flat.

7. Use a jeweler's saw to cut the 18-gauge sterling silver piece close to the size of the pierced silver piece. Sand the surface clean and flat.

VARIATIONS

Linda Darty *To Sandy from Arrowmont*, 2001. 2½ x 1¼ in, (6.4 x 3.2 cm) Sterling silver, enamel. Collection of Sandra Blain. Photo by artist.

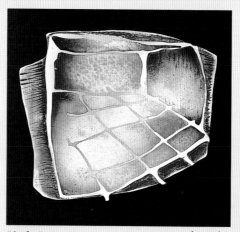

Linda Darty *Tree House Pin #2*, 1982. 2¼ x 1¼ in. (5.7 x 3.2 cm). Sterling silver, enamel. Photo by Dan Bailey.

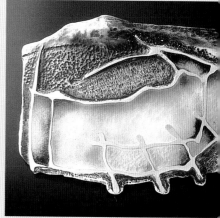

Linda Darty *Tree House Pin #1*, 1982. 2¼ x 1¼ in. (5.7 x 3.2 cm). Sterling silver, enamel. Photo by Dan Bailey.

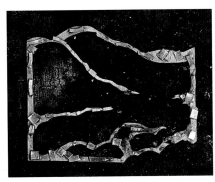

8. Flux the 22-gauge pierced silver piece and the 18-gauge solid silver piece. Carefully apply many hard solder pallions to each line in the pierced design (see photo). Make sure to apply enough solder to cover every part of every line when the metal is heated.

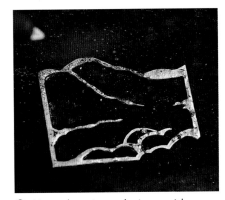

9. Heat the pierced piece with a soft bushy flame until all the solder flows (see photo). Use the torch to follow the flowing line of solder around the piece until it's completely covered. If necessary, add more solder to areas that need it.

10. Once the solder flows, use tweezers to quickly move the pierced piece so it won't stick to the firebrick or soldering block. Heat the fluxed 18-gauge silver piece, and while the metal is still hot, use the tweezers to place the 22-gauge pierced shape on top of it.

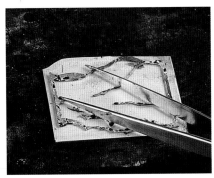

11. When the paste flux turns glassy it essentially "glues" together the two silver pieces. Remove the torch. Use the tweezers to press the 22-gauge pierced silver piece to the 18-gauge piece (see photo). Don't press the metal again when it's very hot, or the thin sawed lines will crack or break. For the same reason, don't hold together the two silver pieces with any type of tweezer setup, such as a third arm.

12. During soldering, it may be necessary to heat the silver elements from the bottom to prevent melting fine design lines. I recommend doing this by raising the lower silver piece with tweezers as shown, rather than by placing the pieces on a screen. A steel screen will act as a heat sink, making it more difficult for you to achieve the correct heating temperature for the silver. Work on charcoal or firebrick for best results.

13. Once the silver starts to color, quickly move the flame to the top surface, heating the piece until the solder line runs completely around it. Follow the solder line with the flame, and be careful not to reticulate the metal.

14. Pickle the silver piece to remove all flux, and then check to see if all areas are completely soldered. If they aren't, simply hammer the silver piece with a leather mallet on a steel block, reflux it, and reheat it until the solder line flows. Do not add solder. If you completely flowed the solder over the 22-gauge pierced silver initially, be confident that you have enough solder now. If you must add solder for some reason, be sure to clean off any residue by using a burr to grind those areas before enameling.

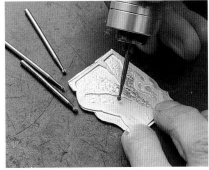

15. If you want to basse-taille texture the surface of the 18-gauge rear metal sheet, do so now. (I create textures using gravers, punches, and tiny round burr bits in a flexible-shaft machine as shown.)

16. Depletion-gild the soldered piece to bring the fine silver to the surface. During this process, be very careful not to overheat the metal and re-flow the solder!

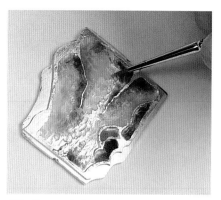

17. Enamel the piece using wet-inlay techniques. To prevent firescale from forming on the sterling silver, cover the entire surface with enamel on the first firing. No counter enamel is necessary.

18. Use diamond sanding sticks to abrade the surface of the piece until the fired enamel and silver are flush. To remove all scratches from the silver, continue to sand the surface with wet emery papers, finishing with the finest grits. Pay particular attention to the silver. (After the final flash firing, any re-sanding of the silver could mar the enamel surface.) Once the silver surface is to your liking, re-fire the enamel until it glosses. When re-fired, the metal will turn black because stoning and sanding took the fine silver layer off its surface. To remove this discoloration, you can use 600-grit sandpaper (it won't disturb the enamel surface).

19. Use a glass brush to thoroughly clean the enamel surface, dry it, and place the brooch upside-down on a trivet. Apply soldering flux to the pin-back findings and to the areas on the back of the brooch where the solder should flow. Carefully position the findings on the brooch, and place a tiny chip of hard solder next to each finding. Use a soft bushy flame to carefully and quickly heat the back of the brooch (see photo). Keep the flame away from the findings so they won't melt. Heat the brooch until you see the solder flow.

20. Let the piece cool before turning it over to examine the surface. When it is cool, you can place the brooch in cold pickle to clean flux from the solder joints. If you're using enamel with low-acid resistance, be sure you don't leave it in the pickle too long.

21. Attach the pin stem and clean up the back of the brooch.

22. Finish the front surface of the brooch in any way you desire. Some of your options include: sanding with wet 600-grit paper, and then rubbing with a soft brass brush and liquid soap; using a buffing wheel for a brighter finish; or darkening the recessed areas of the sterling silver with a liver of sulfur patina.

Plique-à-Jour Earrings

Fay Rooke

Create a luminous pair of pierced plique-à-jour earrings.

Techniques

Using the pierced plique-à-jour enameling technique

Suspending enamel within a pierced cell using capillary attraction

What You Need

Fine silver sheet, 18 gauge

Transparent enamels, medium-fusing, washed and tested

Sifter, 200 mesh

Project kits, page 136

Safety Matters

• Work in a well-ventilated area

What You Do

1. Use a soft-lead pencil to sketch a design for the earrings on tracing paper. The design should have flowing lines and no hard angles or corners. Refine the design by tracing over the initial drawing many times, using fresh paper and a sharpened pencil each time for precision (see photo). Try not to have a line in the drawing that is parallel to the outside edge. Vary the widths of the cell walls and consider whether or not you want to incorporate a hanging device into the design. Keep the

outside contour of the piece simple; circles or ovals are easiest for beginners to make.

2. Blacken the back side of the final tracing paper design with a soft lead pencil to make a "carbon" paper (see photo). You'll use this later to transfer the design to the metal.

3. Wash or sand the fine silver until water sheets off its surface. Paint both sides of the clean metal with white correction fluid or white paint and allow it to thoroughly dry.

4. Place the drawing on the white metal with the blackened side facing the metal surface. Leave sufficient metal around the drawing so you can form a good framework around the design. Tape the design in place or hold it firmly. Use a sharp lead pencil to retrace the drawing and transfer the design to the metal.

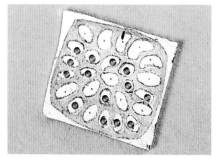

5. Center-punch and then drill small holes in all the cells that will be removed.

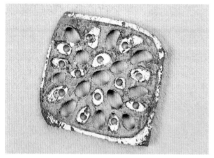

6. Using a jeweler's saw, pierce out every other cell (see photo) and, if necessary, refine the edges of each cell as you go. Saw carefully because the cell walls must be perpendicular to the metal surface. The saw blade should leave a fine, even burr which will help hold in the enamel. Do not remove this sawing burr.

7. Turn the metal over and transfer the other side of the design to the back of the pierced metal. Re-draw or alter the design to accommodate sawing errors and, for clarity, make these changes on the painted metal in a different color ink or pencil. If necessary, re-saw or file the pierced shapes, making sure to leave an adequate wall between the cells. Once the filing is complete, finish sawing around the outside of the design.

8. Gently clean the pierced metal using a glass brush and soapy water. Be careful to leave the burr around the edges of the cells. Rinse well. (Keep the glass brush you use with the soapy water separate from the one you use to clean enamels between firings. If soapy water contaminated a crack in an enamel, cloudiness would be present in the final firing. Save this soapy glass brush for use on metal surfaces only.) Pat the piece dry and handle it by the edges. The cell walls should be perfect with no jags as these will be magnified in the finished piece once the enamel is inlayed. (It may be helpful to check the cell edges on a light table.)

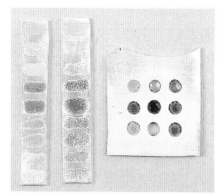

9. Select medium-fusing transparent enamels that fire bright and clear. Pre-sift small amounts with a 200-mesh sifter. Wash the 200-mesh colors well. Make a test strip of the

enamels you want to use by applying each color separately to a piece of clean 18-gauge fine silver. Dry and fire the test strip until the softest enamel begins to gloss, and then quickly remove the strip from the heat (see photo). This test indicates the lowest to highest fusing range. Enamels that are too soft will sag or burn through quickly. Hard enamels will be difficult to fuse without losing some other colors. Make a second test strip and fire the same colors to maturity. Depending on the fusing temperatures, you may want to choose different colors that are more compatible.

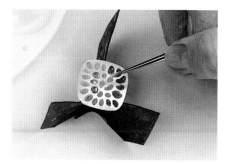

10. Put the pierced piece on a clean trivet with filed edges or, preferably, hold it by its edges. Either paint the whole piece with an undiluted holding agent before applying wet enamel, or add a little binder to the wet enamel. Begin wet-inlaying the washed enamel. Using a medium-size clean paint brush, pick up the enamel and apply it using a sweeping, fluid motion. The enamel will adhere to edges of the cell like a suspended drop. Use the same fluid motion and sweep the suspended drop of enamel from the center of the cell out to the cell edge. Capillary action will cause the enamel to "jump" to the edge and flow around the cell rim, making a concave pool of enamel that is thicker at the wall edge. Use a lint-free tissue or towel to wick the moisture from each cell as you proceed, being careful that moisture does not transfer between the cells. Working from the center to the outside, complete all cells; then remove stray enamel grains from the metal surface.

11. Check the cells by holding the piece to the light. Note that some cells have a thinner application of enamel, and they will burn out first.

12. Fill these cells with additional enamel and let the piece thoroughly dry. Turn the piece upside down on the trivet and briefly fire at approximately 1400° F (760° C) until the enamel is shiny but still orange-peeled. Carefully watch the piece. Depending on your kiln, this firing may take only 45 to 60 seconds.

VARIATION

Fay Rooke *Earrings*, 2002. 2 x 4 cm. Fine silver, enamel; embossed copper foil plique-à-jour. Photo by artist.

(Noting how the piece fired, adjust your kiln temperature if necessary.)

13. Repeat the inlay-and-firing procedure, always thinly applying the enamel, turning the piece over each time, and applying the enamel where the glass is convex. Continue until all cells are full and fired glossy.

14. Use wet/dry emery papers in a circular motion under running water to remove excess enamel until the plique a jour surface is level on both sides. Refine the earring edge by filing on an angle to make a tapered edge, and then sand smooth. Completely finish the metal surface, polishing it to your satisfaction with the fine wet/dry papers. Gently glass brush the piece under running water.

15. Place the enameled earrings front-side-up for the final firing. (Be sure that the front sides of the earrings are complimentary.) Fire at the same temperature as before, but for a slightly longer time until the enamel surface is smooth and glossy. The cells will have a slightly concave surface creating "pools of light", and a maximum reflecting surface.

16. Insert handmade or commercial ear wires into the holes in the framework to finish the piece.

HOT TIPS

—Plique-à-jour uses a minimum amount of enamel so wash small amounts.

—Though you could use a sheet of mica as a backing rather than using the capillary action technique, a bit of the mica does fire into the glass surface. This would require abrasion techniques to completely remove the mica. The capillary action technique produces more brilliant clear colors and lessens the chances of cracking the piece during sanding and stoning.

—The thicker the metal you pierce, the thicker the enamel will be, resulting in more opportunity to apply color and create light reflection. Thicker metal does, however, require more precise sawing.

—Smaller cell openings are easier to fill using the capillary action technique. Cells with sharp angles will be likely to crack as the metal compresses on the glass in the angle.

Enamel & Encaustic Assemblage

Jane Harrison

Include enamel in a striking mixed-media wall piece.

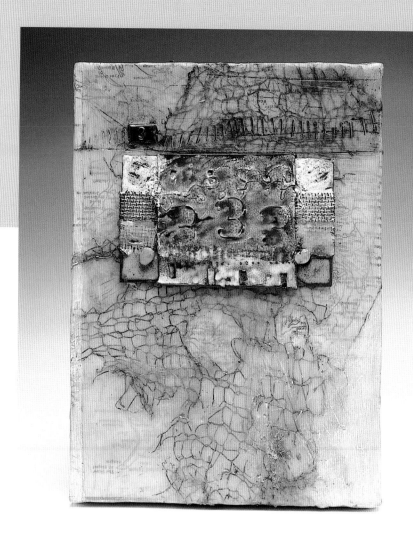

Techniques

Enameling on copper foil

Working with basse taille surfaces, and watercolor and acrylic enamels

Combining enameling with encaustic and painting media

What You Need

Copper foil, approximately 2 x 3 inches (5 x 7.6 cm)

Copper mesh scrub pad

Copper screen, small piece

Small copper nails

Birch plywood, 5 x 7 x ½ inch (12.7 x 17.8 x 1.3 cm)

Liquid enamels, white and clear

Powdered enamel, transparent

Acrylic enamels

Cellulose-based glue

Ledger paper or other old paper

Paraffin

Oil paints

Glass etching cream (optional)

Found objects (optional)

Thread (optional)

Dental tool

Stapler and staples

Electric skillet for wax

Heat gun

Dressmaker's tracing wheel (optional)

Project kits, page 136

Safety Matters

• Wear a respirator when sifting enamel

• Use ventilation when enameling, soldering, and pickling

What You Do

1. Mark and drill holes in the copper foil so the finished enamel can later be attached to the plywood backing.

2. Anneal the copper foil in the kiln or with a torch, and then clean the metal by putting it in pickle to remove as much firescale as desired. Punch designs and chase textures into the softened foil using tools such as stamps or nails. (For this project, rectangular and circular hole punches were used to make rows of small holes in the foil. A dressmaker's tracing wheel was used to make lines.)

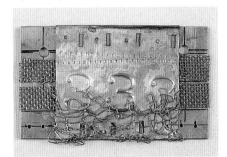

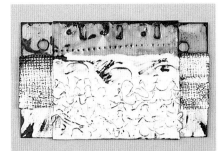

3. Assemble a copper collage by simply crimping some strips of copper mesh and a flattened copper scrub pad around the imprinted foil.

4. Paint liquid white and liquid clear enamel on selected areas of the metal collage. Let the enamel dry.

5. Over-fire the metal collage until copper oxide creates green areas in the enamel.

6. Paint gum binder on the center panel of the collage, sift transparent enamel onto the binder

(in this case, turquoise), and let dry. Use a dental tool to scrape lines and dots into the enamel. Fire the collage to the fusing temperature (see photo).

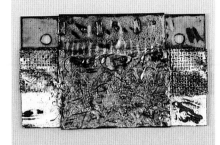

7. Paint acrylic enamel in selected areas on the collage, and fire the piece to fusing temperature. (Here, brownish-red acrylic enamel was used.)

8. Paint between the strips of copper mesh and over the top margin of the center panel. Slightly smear the paint a little with fingers if

VARIATIONS

Jane Harrison Untitled, 2003. 7 x 5 in. (17.8 x 10.2 cm). Copper, enamel, wood, encaustic, found objects, oil paint. Photo by Linda Darty.

Jane Harrison Untitled, 2003. 7 x 5 in. (17.8 x 10.2 cm). Copper, enamel, wood, encaustic, found objects, oil paint. Photo by Linda Darty.

Jane Harrison Untitled, 2003. 7 x 5 in. (17.8 x 10.2 cm). Copper, enamel, wood, encaustic, found objects, oil paint. Photo by Linda Darty.

desired. Let the enamel dry, and fire it to fusing temperature. (On this piece a very light blue-violet acrylic enamel was used.)

9. Once you are pleased with the colors, you can matte the finished enamel surface with glass etching cream if you wish.

10. Sand the edges of the birch plywood. Using a cellulose-based glue, completely cover the board (including its edges) with an old piece of ledger paper. (Since you'll be using wax over the paper, it's necessary that no acrylic or polymer materials be used to adhere it to the wood.) Let the glue dry.

11. Melt paraffin in an electric skillet reserved for that purpose. With a paintbrush, apply several coats of clear melted wax to the surface of the papered wood.

12. Wrap a second flattened copper scrub pad around the top section of the board and staple it securely to the back (see photo). Add a line of thread across the top of the piece, and tie it to a small copper nail on each edge.

13. With a paintbrush, apply several coats of clear melted paraffin over the surface of the covered wood as shown. Using a heat gun, slightly melt the wax in various places to assure it's fused to the paper, scrub pad, and wood.

14. Attach the enamel to the encaustic-covered panel with copper nails. Wax the nail heads to make them less shiny.

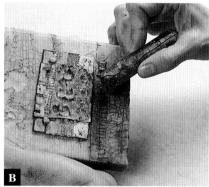

15. You can scrape the wax surface smooth and/or use small nails to create designs in selected areas (photo A). The wax is malleable, so you can use a tracing wheel to emboss some lines into the wax; or hammer copper scrub pad pieces into the wax, and then remove them to create texture. After the waxed piece is completely cool, use oil paints to stain the surface (photo B). You also can hammer found objects onto the piece if desired.

The Brown and Sharpe (B & S) Gauge for Sheet Metal

GAUGE NUMBER	THICKNESS in inches	in millimeters
3/0	.409	10.388
2/0	.364	9.24
1/0	.324	8.23
1	.289	7.338
2	.257	6.527
3	.229	5..808
4	.204	5.18
5	.181	4.59
6	.162	4.11
7	.144	3.66
8	.128	3.24
9	.114	2.89
10	.101	2.565
11	.090	2.28
12	.080	2.03
13	.071	1.79
14	.064	1.625
15	.057	1.447
16	.050	1.27
17	.045	1.114
18	.040	1.016
19	.035	.889
20	.031	.787
21	.028	.711
22	.025	.635
23	.022	.558
24	.020	.508
25	.017	.431
26	.015	.381
27	.014	.376
28	.012	.304
29	.011	.29
30	.01	.254
31	.008	.203
32	.0079	.199
33	.007	.177
34	.006	.152
35	.0055	.142
36	.005	.127

Select Bibliography

Andrew I. Andrews. "Proceedings of the First Annual Porcelain Enamel Institute Forum." Urbana, Illinois: University of Illinois, 1937.

Ball, Fred. *Experimental Techniques in Enameling*. New York: Van Nostrand Reinhold Company, 1972.

Barsali, Isa Belli. *European Enamels*. New York: The Hamlyn Publishing Group, Ltd., 1969.

Bates, Kenneth F. *Enameling Principles and Practice*. Cleveland and New York: The World Publishing Company, 1951.

Bates, Kenneth. *Enamelist*. New York: The World Publishing Company, 1967.

Carpenter, Woodrow, Bill Helwig, and Tom Ellis.*Thompson Enamel Workbook*. Newport, Kentucky: Thompson Enamel, 1997.

Cellini, Benvenuto. *The Treatises of Benvenuto Cellini on Goldsmithing and Sculpture*. Translated by C.R. Ashbee. New York: Dover Publications, 1996.

Cunynghame, Henry. *On the Theory and Practice of Art: Enameling Upon Metals*. Whitehall Gardens: Archibald Constable and Company,1899.

De Koningh, H. *The Preparation of Precious and Other Metal Work for Enameling*. New York: The Norman W. Henley Publishing Co., 1930.

Dossie, Robert. *"The Handmaid to the Arts"* in *Ordinary to His Majesty*. London: J. Nourse. Reprint, New York: Dover Publications, 1967.

Garner, Sir Harry. *Chinese and Japanese Cloisonne Enamels*. London: Faber and Faber, Ltd., 1976.

Glass on Metal. Bellevue, Kentucky: Thompson Enamel, Inc. 1982–2004.

Harper, William. *Enameling, Step-by-Step*. New York: Golden Books, 1973.

Larom, Mary. *Enameling for Fun and Profit*. New York: David McKay Company, 1954.

McCreight, Tim. *The Complete Metalsmith: An Illustrated Handbook*. New York: Sterling Publishing, 1991.

McCreight, Tim. *Working with Precious Metal Clay*. Portland, Maine: Brynmorgen Press, 2000.

Metalsmith. Lisle, Illinois: The Society of North American Goldsmiths, 1980–2004.

Speel, Erika. *Dictionary of Enameling*. Hampshire, United Kingdom: Ashgate Publishing Company, 1998.

Sullivan, Edward J. and Ruth Krueger Meyer. *The Taft Museum European Decorative Arts*. New York: Hudson Hills Press, Inc., 1995.

Theophilus. *On Divers Arts*. Translated by John G. Hawthorne and Cyril Stanley Smith. New York: Dover Publications, 1979.

Thompson, Thomas E. *Enameling on Copper and Other Metals*. Highland Park, Illinois: Thomas C. Thompson Company, 1950.

Untracht, Oppi. *Enameling on Metal*. Radnor, Pennsylvania: Chilton/Haynes, 1972.

Vogelzang, V. and E. Welch. *Graniteware*. Des Moines, Iowa: Wallace-Homestead Book Company, 1981.

Winter, Edward. *Enamel Art on Metals*. New York: Watson-Guptill Publications, 1958.

Acknowledgments

Writing this book has been a true collaboration. I'd like to acknowledge the generous support of many friends, colleagues, and especially the faculty and students I work with at East Carolina University. Their enthusiasm for this project ignited my own, and their patient assistance was invaluable.

I'm especially grateful to the project designers: **Robert Ebendorf**, **Tim Lazure**, and **Mi Sook Hur** are professors and my colleagues in the East Carolina University School of Art and Design metals program, and excellent metalsmiths willing to try their hand at enameling! **Sharon Massey**, **Kathryn Osgood**, **Kate Cathey**, and **Adrienne Grafton** are outstanding graduate students in the metals program at ECU. **Jane Harrison**, who recently received her MFA in painting from ECU, inspired us all when she took just one enameling class. **Jennifer Mokren** is an enamellist and professor in metals at the University of Wisconsin at Greenbay, Wisconsin. **Fay Rooke** is an internationally recognized enamellist from Burlington, Ontario, Canada, who founded and instructed in the enameling program at the Ontario College of Art and Design in Toronto.

In addition to the many enamellists who sent gallery images, special thanks goes to the artists listed here who contributed demonstration samples so that I could more clearly share their personal research.

Jesse Bert	**Sharon Massey**
Beth Blake	**Christina Miller**
Whitney Boone	**Barbara Minor**
Harlan Butt	**Nick Mowers**
John Cogswell	**Matt Owen**
Gina Cox	**Sarah Perkins**
Kathleen Doyle	**Suzanne Pugh**
Scott Eagle	**Thomas Reardon**
Ray Elmore	**Dindy Reich**
Corey Fong	**Mary Reynolds**
Annie Grimes	**Eva Roberts**
Jennifer Hatlestad	**Ken Rockwell**
Lindsey Hardin	**Fay Rooke**
Paul Hartley	**Nate Sealy**
Bill Helwig	**Coral Shaffer**
Barbara Hopkins	**Barbara Simon**
Melissa Huff	**Mary Steenburger**
Mi Sook Hur	**Jean Tudor**
Mickey Johnston	**Elizabeth Turrell**
Kimberly Keyworth	**Jessica Turrell**
Rebekah Laskin	**Michael Voors**
Joan MacKarell	**Catherine Walker**
Melissa Manley	**Melissa Walter**
Joan Mansfield	**Kathleen Wilcox**

I want to thank Penland School of Crafts for allowing me to shoot the demonstration photos in the same studio in which I learned to enamel 28 years ago! I'm grateful to all the staff at Lark Books, and especially to **Deborah Morgenthal** for her expertise and patience, **Marthe Le Van** for her time, **Rebecca Guthrie** for graciously and skillfully organizing the many slides with their captions, and **Kathy Holmes**, for giving me confidence with her keen eye and artistic abilities. I sincerely thank **Carol Taylor** and **Rob Pulleyn** for patiently supporting me.

I'm especially indebted to my mentor, **Bill Helwig**, not only for his very careful and constructive reading of the manuscript, but also for being my best enameling teacher and good friend. I'm deeply grateful to **Woodrow Carpenter** for his assistance as I researched and studied both the practice and history of enameling. His generosity and patience are as remarkable as the depth and breadth of his knowledge. I thank **Woodrow Carpenter**, **Erika Speel**, **Bill Helwig**, and the editors of *Glass on Metal*, for allowing me to include information in the text, originally written by them and published in that journal. I want to thank **Nancy Levine** and **Jean Tudor** for their advice and assistance with my historical research and **Elizabeth Turrell**, **Bernard Jazzar**, **Robert Corson**, **Susan Willoughby**, **Bill Griffith**, **Gretchen Goss**, **Kathleen Brown**, **Jacqueline Vittimberga**, **Nancy Worden**, and **Dr. Panicos E. Michaelides** for helping me obtain images of historical enamel works. **Hillary Dorsky's** review of the manuscript was most appreciated, and it led to many new tips and further clarification of the text. **Kathleen Doyle** and **Paula Garrett** came to my aid during the most difficult of days, filling me with gratitude for their treasured friendship. Thanks to **Paulus Berensohn** for his thoughtful words, and to all of my Penland family for shepherding me through a snowy winter of writing and for the friendships and summer memories that gave me strength through long working days.

I'm forever grateful to **Bill** and **Jane Brown** for giving me the opportunity to live and work at Penland School of Crafts in the 1970s, and for opening the door to the many experiences that have continued to shape my life and inspire my passion.

Finally, I'm deeply grateful to my family; my mother and father for encouraging my artistic endeavors; and **Terry**, **Warren**, and **Elliot** whose heartfelt support is treasured more than words can say.

I've written this book with fondest memories of **John Satterfield**, my teacher, dear friend, and former colleague, who gave me not only my metalworking skills, but also the confidence to teach. I dedicate the book to **Warren**, **Elliot**, **Brady**, the **summer cousins**, and **all my students**, with bright hopes and dreams that your futures, too, may be filled with a passion; and to **Paula**, who helped me find the artist in my soul.

Contributing Artists

Index